Engineering Reliability

ASA-SIAM Series on Statistics and Applied Probability

The ASA-SIAM Series on Statistics and Applied Probability is published jointly by the American Statistical Association and the Society for Industrial and Applied Mathematics. The series consists of a broad spectrum of books on topics in statistics and applied probability. The purpose of the series is to provide inexpensive, quality publications of interest to the intersecting membership of the two societies.

Editorial Board

Barlow, R. E., *Engineering Reliability*

Czitrom, V. and Spagon, P. D., *Statistical Case Studies for Industrial Process Improvement*

Engineering Reliability

Richard E. Barlow
University of California, Berkeley
Berkeley, California

Society for Industrial and Applied Mathematics
Philadelphia, Pennsylvania

ASA

American Statistical Association
Alexandria, Virginia

Library of Congress Cataloging-in-Publication Data

Barlow, Richard E.
    Engineering reliability / Richard E. Barlow.
        p. cm. -- (ASA-SIAM series on statistics and applied probability)
    Includes bibliographical references and index.
    ISBN 0-89871-405-2 (pbk.)
    1. Reliability (Engineering).   I. Title.   II. Series.
    TA169.B36   1998
    620'.004'52--dc21   97-44019

This work is dedicated to my wife, Barbara.

Contents

*Chapters and sections marked with asterisks discuss special topics.

Preface

Engineering reliability concerns failure data analysis, the economics of maintenance policies, and system reliability. The purpose of this book is to develop the use of probability in engineering reliability and maintenance problems. We use probability models in

1) *analysis* of failure data,
2) *decision* relative to planned maintenance,
3) *prediction* relative to preliminary design.

Engineering applications are emphasized and are used to motivate the methodology presented.

The background required is at least a first course in probability. Although the probability calculus will be reviewed, it would be best to have had previous exposure. Some previous knowledge of engineering or physics at the beginning level would also be helpful. The academic level required is upper division undergraduate or first year graduate, but no previous course in statistics or measure theory is needed.

We will make extensive use of spreadsheets in the exercises. Spreadsheets such as EXCEL are now ubiquitous and contain most of the mathematical and statistical functions needed to solve many of the exercises in this book. They are also very useful for the graphical analysis of failure data contained in the book.

Part I is devoted to the analysis of failure data, particularly lifetime data and failure counts. We begin by using a new approach to probability applications. The approach starts with finite populations and derives conditional probability models based on engineering and economic judgments. The infinite population conditional probability models that are most often used are approximations to these finite population models. The derived conditional probability models are then the basis for likelihood functions useful for the analysis of failure data.

Part II addresses the economics of maintenance decisions. We begin with the economics of replacement decisions using a deterministic approach. Next we discuss Deming's ALL or NONE rule for inspection sampling. Optimal age replacement policies are considered with emphasis on the time value of money and discounting.

Part III focuses on system reliability. We begin with efficient algorithms for computing network reliability. Networks or "block diagrams" are abstract system representations useful for both reliability prediction and maintenance optimization. Fault tree analysis as presented in Chapter 7 is one of the most useful tools in identifying system failure scenarios. Chapters 6, 7, and 8 concentrate on network and system reliability, availability, and maintainability.

The Bayesian approach is described in the introduction. It might be wise to cover this material before beginning Chapter 1. Part I and Chapters 9 and 10 of Part III use the Bayesian approach. The probabilistic foundation for influence diagrams is presented in Chapter 9. Fault trees are special cases of probabilistic influence diagrams. Influence diagrams are alternatives to decision trees and as such are useful in decision making. Chapter 10 presents a tutorial introduction to decision influence diagram construction and solution. Appendix A discusses the inconsistency of classical statistics, while Appendix B provides an axiomatic argument due to H. Rubin for the consistency of the Bayesian approach to decision analysis. The last section in each chapter consists of notes and references. Chapters marked with asterisks discuss specialized topics, and starred exercises are mathematically challenging.

Some other related textbooks on system reliability are I. B. Gertsbakh (1989), A. Høyland and M. Rausand (1994), R. D. Leitch (1995), R. E. Barlow and F. Proschan (1981), and R. E. Barlow and F. Proschan (1996).

See `http://www.ieor.berkeley.edu/~ieor265` regarding errata, answers to frequently asked questions, and links to relevant information sources.

Richard E. Barlow
Berkeley, CA

Acknowledgments

I would like to thank the Division of Statistics at the University of California at Davis for the opportunity to develop this material in the summer of 1996. Research for this book was partially supported by U.S. Army Research Office contract DAAG29-85-K-0208 and Air Force Office of Scientific Research grant AFSOR#F49620-93-1-0011 with the University of California.

Introduction

Engineering reliability is (or should be) failure oriented. The problem is to predict when or if failure will occur when a device is used. This information can then be used to determine inspection and maintenance policies as well as warranties. It can also be used to predict costs due to maintenance and eventual failure if failure occurs while the device is in operation.

The commonly used definition of engineering reliability is "*Reliability* is the *probability* of a device performing its purpose adequately for the period of time intended under operating conditions encountered."

Historical perspectives on reliability theory. The mathematical theory of reliability has grown out of the demands of modern technology and particularly out of the experiences with complex military systems. One of the first areas of reliability to be approached with any mathematical sophistication was the area of machine maintenance. Queuing and renewal theory techniques were used early on to solve problems involving repair and inspection.

In 1939 Walodie Weibull, then a professor at the Royal Institute of Technology in Sweden, proposed the distribution, later named after him, as an appropriate distribution to describe the breaking strength of materials. In Chapter 4 we discuss Weibull's theory and propose new methods for analyzing the strength of materials.

In 1951 Epstein and Sobel began work on the exponential distribution as a probability model for the study of item lifetimes [see Epstein and Sobel (1953)]. This probability model, as well as most others, is based on the concept of an infinite or unbounded population size. In Chapter 1 we derive the finite population version of this probability model. Only in the limit, as population size approaches infinity, do we obtain the properties characteristic of the exponential distribution. Among these properties is the memoryless property or nonaging property. Although this property would seem to limit the usefulness of the exponential distribution for lifetime, it has continued to play a critical role in reliability calculations. A fundamental reason for the popularity of the exponential distribution and its widespread exploitation in reliability work is that it leads to simple addition of failure rates and makes it possible to compile design data in a simple form.

Research on coherent system structure functions (or general system reliability) began with the 1961 paper by Birnbaum, Esary, and Saunders. It put the previous work by Moore–Shannon on super-reliable relays in a more abstract setting. The connection between coherent structures and the class of life probability distributions containing the exponential distributions was discovered by Birnbaum, Esary, and Marshall (1966).

The engineering reliability emphasis in the 1970s was directed to fault tree analysis, which we discuss in Chapter 7. This was and still is motivated by nuclear-power reactor safety considerations among other engineering safety problems.

In the 1980s, a great deal of reliability work was concerned with network reliability. This was motivated by the early Advanced Research Projects Agency (ARPA) network, the forerunner of today's Internet and world wide web. In Chapter 6 we discuss some of the most useful work in this area. Although the problem of computing the probability that distinguished nodes in a network are connected is an NP-hard problem, efficient algorithms do exist for special classes of networks.

In the 1990s there has been reliability research in new directions led by M. B. Mendel. The motivation for this research is based on the belief that most sample space representations that have been considered in engineering statistics are not proper Euclidean spaces. Therefore, taking inspiration from physics, an approach to problems in engineering statistics based on differential geometry is needed. This point of view has led to a number of recent publications concerning engineering reliability problems, including Shortle and Mendel (1994) and (1996).

What is probability? We defined reliability as the probability of a device performing its purpose adequately for the period of time intended under operating conditions encountered. But what is probability?

For the physicist, probability is defined as follows: "By the *probability* of a particular outcome of an observation we mean our estimate for the most likely fraction of a number of repeated observations that will yield that particular outcome" [Feynman,* *The Feynman Lectures on Physics*, Vol. I]. The physicist's idea is to imagine repeating an experiment which some of the time produces an event, say, A, and calculating $\frac{N_A}{N}$ and then setting

$$P(A) = \frac{N_A}{N},$$

where N is the number of experiments and N_A is the number of times the event A occurs. Note that this definition requires repetition. There are at least two snags to this idea, namely, the following.

(1) No event is exactly repeatable. At the very least, the occurrence times are different.

(2) Our probability may change as our knowledge changes.

---

*Feynman notes in Vol. I, p. 6-2 another rather "subjective" aspect of his definition of probability.

Nevertheless, it is useful to use this frequency idea of probability in many physical applications. We will see later that this idea breaks down in many cases (in quantum mechanics, for example).

Another definition of probability is that "Probability is a degree of belief held by an analyst or observer." The idea here is that probability is an *analytical tool based on judgment useful for making decisions.* Of course we are free to adopt observed frequencies as our probabilities if we are consistent, but observed frequencies are not otherwise probabilities according to this definition. Using either definition, probability P should satisfy certain computational "laws":

(1) **Convexity law.** If E is an event, then the probability of E, $P(E)$, satisfies

$$0 \leq P(E) \leq 1.$$

The reason for the name is that we will use convex combinations of probability weights in computing expectations.

(2) **Addition law.** If E_1 and E_2 are any two mutually exclusive events, then

$$P(E_1 \text{ or } E_2) = P(E_1) + P(E_2).$$

This can be extended to any n mutually exclusive events.

(3) **Multiplication law.** If E_1 and E_2 are any two events, then

$$P(E_1 \text{ and } E_2) = P(E_1 \mid E_2)P(E_2) = P(E_2 \mid E_1)P(E_1).$$

This too can be extended to include any n events. The event to the right of the vertical bar is either given or assumed.

When writing $P(E_1 \mid E_2)$, we mean the probability of E_1 *were* E_2 known. It is important to use the subjunctive tense to assess the conditional probability, since it is not necessary that E_2 be known. Also it is important to emphasize that $P(\bullet \mid \bullet)$ is a *probability function* of the first argument but not the second.

Bayes' Formula.

Using the multiplication law, we can write

$$P(E_2 \mid E_1) = \frac{P(E_1 \mid E_2)P(E_2)}{P(E_1)}.$$

This is Bayes' formula in its most simple form. If event E_1 is observed and we wish to calculate the conditional probability of an event E_2 that is unknown, then we can use the above formula to make this inference. $P(E_2)$ in the formula is our *prior* probability for the event E_2. $P(E_1 \mid E_2)$ is the conditional probability of E_1 conditional on E_2. However, since we suppose that E_1 has been observed, $P(E_1 \mid E_2)$ is called the *likelihood* of E_2 given the data E_1. It is important to remark that the likelihood is *not* a probability function of its second argument E_2 but simply a nonnegative real valued function. $P(E_2 \mid E_1)$ is called the *posterior* probability of E_2 conditional on E_1.

The Bayesian approach.

Historically, the XVIII and XIX century Bayesian paradigm was based on Bayes' theorem and reference priors. It was thought (incorrectly) that complete ignorance could be modeled by "noninformative" priors. R. A. Fisher in the 1920s pointed out that if a uniform prior for $\pi(0 \leq \pi \leq 1)$ expresses ignorance, then the induced probability for π^2 should also be uniform since π^2 is just as uncertain as π. But the induced probability distribution for π^2 is not uniform.

There is no logically valid way to model "complete ignorance" nor is it desirable to try to do so. The use of so-called flat or constant prior probability functions can, however, be justified in certain situations based on the analyst's judgment of what he/she knows relative to the "standardized likelihood." The standardized likelihood is the likelihood multiplied by a constant to make it sum or integrate to one as a function of its second argument. In this situation of relative ignorance, the standardized likelihood and posterior density are sufficiently close to be interchangeable. This is called the "principle of precise estimation" in DeGroot (1970) and, in general, the principle is not dependent on sample size.

Bayesian statistics in the twentieth century has been greatly influenced by the ideas of de Finetti (1937). In his view, probability is based on personal judgment. But it is not arbitrary! The laws of probability must be obeyed, and various logical ways of thinking about uncertainty must be considered. In particular, the classical idea of randomness is replaced by the judgment of exchangeability. This judgment leads to de Finetti's famous representation theorem. Random quantities $X_1, X_2, \ldots, X_n$ are exchangeable for you if in your opinion their joint distribution is invariant under permutations of coordinates, i.e.,

$$F(x_1, x_2, \ldots, x_n) = F(x_{\pi_1}, x_{\pi_2}, \ldots, x_{\pi_n}), \tag{1}$$

where $F(\bullet, \bullet, \ldots, \bullet)$ is a generic joint cumulative distribution for $X_1, X_2, \ldots, X_n$ and equality holds for all permutation vectors $(\pi_1, \pi_2, \ldots, \pi_n)$ and vectors $(x_1, x_2, \ldots, x_n)$. For example, suppose a set of items were exchangeable for you relative to failure. Then, for you, their marginal probabilities of failure would be the same and, more generally, equation (1) would hold. If binary (0 or 1) random quantities or events are judged exchangeable and there is a conceptually infinite number of them, then the probability that k-out-of-n are 1 is necessarily of the form

$$P\left(\sum_{i=1}^{n} x_i = k\right) = \binom{n}{k} \int_0^1 \pi^k (1 - \pi)^{n-k} p(\pi) d\pi \tag{2}$$

for some density $p(\bullet)$ (or, more generally, a distribution). Bayesians would interpret $p(\bullet)$ as a prior for π. The parameter π could be thought of as a "propensity" or "chance" and is related to the long-run ratio of occurrences to trials if such an infinite sequence of trials could be actually carried out. Given data, namely, x occurrences in N trials, the conditional probability of an additional k occurrences in an additional n trials looks like (2) with $p(\pi)$ replaced by the posterior density for π, namely, $p(\pi \mid x, N)$, calculated via Bayes' formula using densities.

FIG. 1. *The two-slit experiment.*

The Bayesian approach is characterized by the idea that all probabilities, likelihood models included, are based on judgment. To quote de Finetti, "Probabilistic reasoning—always to be understood as subjective—merely stems from our being uncertain about something. It makes no difference whether the uncertainty relates to an unforeseeable future, or to an unnoticed past, or to a past doubtfully reported or forgotten; it may even relate to something more or less knowable (by means of a computation, a logical deduction, etc.) but for which we are not willing to make the effort; and so on."

In Einstein's theory of relativity, mass, velocity, and time are relative to the observer. The same is true for probability; i.e., probabilities are relative to the observer or analyst. Also, probabilities are conditional, conditional on what is known. For this reason, conditional probabilities play a central role in this approach.

The role of the observer (or analyst) and the second law. The three laws of probability considered earlier can actually be "proved" from more fundamental principles [cf. Lindley (1985)]. Feynman's conventional definition of probability satisfies these three laws *except* in the case of quantum mechanics. The second law, the addition law, holds in this case only if an observer could actually "observe" the objects in question. In the experiment described below, those objects are electrons.

To understand this statement, consider the famous two-slit experiment of quantum mechanics [Feynman, Vol. 1, Chapter 37]. Imagine an electron gun shooting off electrons through a slit in a barrier as in Figure 1. To the right is another barrier with two slits. Finally, there is another track to the right of this with an electron detector. The detector counts the number of electrons arriving at a point a distance x from some origin. For each position x we calculate the ratio of the number arriving at x in a specified time interval to the number traveling through the initial slit. Let $p(x)$ be this ratio for each x.

If a detected electron is like a particle, then it has either gone through the first slit or the second slit but not both. If we close slit 2, then we get the ratio $p_1(x)$ for the frequency of electrons hitting position x after traveling through the first slit. Likewise if we close slit 1, we get a corresponding frequency $p_2(x)$. Now open both slits. If the electron behaves like a particle, we should find that $p(x) = p_1(x) + p_2(x)$ since this is the addition law for mutually exclusive events. However, this is not at all what is detected. Figure 2 shows what is actually detected.

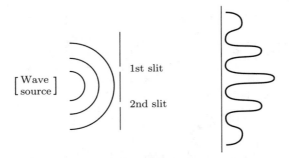

FIG. 2. *Observed frequencies.*

Now suppose a light source is used to "see" each electron as it leaves the original slit and is used to determine which of the two slits to the right that it "goes through." In this case, when each electron event is observed at x, the addition law does hold! These are experimental results that we are describing.

When we do not observe each electron's trajectory but only count frequencies in a kind of objective manner, the laws of probability annunciated above *do not hold.* When there is included in the experiment an observer who actually observes each electron, his frequencies at positions x now satisfy the addition law of probability. When we look, the distribution is different from that when we don't look!

We may adopt a frequency as our probability, but it is a probability only after we adopt it. Probability must be considered a degree of belief relative to some observer or analyst even when we have large numbers of apparently repetitive events. However, a consensus on probabilities is sometimes possible, and that is usually what we desire. What is called "objectivity" is perhaps better described as "consensus."

PART I

Failure Data Analysis

CHAPTER **1**

The Finite Population Exponential Model

1.1. Finite populations.

Suppose we have exactly N items $\{1, 2, \ldots, N\}$. That is all there are. For example, we could produce exactly N items to a "special order." For another example, there were only five space shuttles in 1996. All populations are actually finite, although it is often useful to consider unbounded populations. It is usual to start with the assumption that random quantities (also called random variables) are *independent, identically distributed* (*iid*). However, this cannot be justified for finite populations.

We are interested in the unknown lifetimes $(x_1, x_2, \ldots, x_N)$ of these items. To learn something about these lifetimes we may be willing to sacrifice say, n, of them. Having observed $(x_1, x_2, \ldots, x_n)$ we may wish to infer something about the population average lifetime or to predict the remaining lifetimes.

Suppose the parameter of interest to us is the average lifetime of all N such items. Let

$$\theta = \frac{\sum_{i=1}^{N} x_i}{N}$$

be this average lifetime. Let $\boldsymbol{x}_N = (x_1, x_2, \ldots, x_N)$ and suppose we are indifferent to vectors of size N of such possible lifetimes *conditional on* θ; i.e., our generic joint probability function p for such vectors will satisfy

$$p(\mathbf{x}_N) = p(\mathbf{y}_N)$$

when $\sum_{i=1}^{N} x_i = \sum_{i=1}^{N} y_i = N\theta$. Since lifetime distributions will be considered absolutely continuous in this and following sections, $p(\bullet)$ will be a joint probability density function. It is not a probability. We will use $P(\bullet)$ when referring to the probability of events.

Note that when $N = 2$, (x_1, x_2) is restricted to the line segment as illustrated in Figure 1.1.1.

Because $p(x_1, x_2)$ is constant on the manifold $\{x_1, x_2 \mid x_i \geq 0, x_1 + x_2 = 2\theta\}$ and $x_2 = 2\theta - x_1$, it follows that $p(x_1 \mid x_1 + x_2 = 2\theta)$ is a constant and this

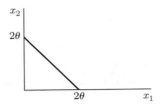

FIG. 1.1.1. *Simplex manifold when $N = 2$.*

constant must be $\frac{1}{2\theta}$. In general, we can show that

$$
p\left(x_1, x_2, \ldots, x_n \mid x_i \geq 0, \sum_{i=1}^{N} x_i = N\theta\right)
$$
$$
= \frac{(N-1)}{N\theta} \frac{(N-2)}{N\theta} \cdots \frac{(N-n)}{N\theta} \left[1 - \frac{\sum_{i=1}^{n} x_i}{N\theta}\right]^{N-n-1} \tag{1.1.1}
$$

for $0 \leq \sum_{i=1}^{n} x_i \leq N\theta$. The joint density is 0 elsewhere. The density is only valid for $n = 1, 2, \ldots, N-1$. The density does not exist for $n = N$.

Fix n and let $N \uparrow \infty$ so that

$$
\lim_{N \uparrow \infty} p\left(x_1, x_2, \ldots, x_n \mid x_i \geq 0, \sum_{i=1}^{N} x_i = N\theta\right) = \prod_{i=1}^{n} \left[\frac{1}{\theta} e^{-\frac{x_i}{\theta}}\right] \tag{1.1.2}
$$

since $\lim_{N \uparrow \infty}(1 + \frac{1}{N})^N = e$, i.e., Euler's "e."[1] Note that we have *conditional independence* (conditional on θ) only in the limit as the population size approaches infinity.

THEOREM 1.1.1. *Let N be the population size. Fix $\theta > 0$ and assume that we are indifferent to points on the manifold*

$$
\left\{\mathbf{x}_N \mid x_i \geq 0, \sum_{i=1}^{N} x_i = N\theta\right\}
$$

in the sense that, conditional on θ, our joint probability function $p(\bullet)$ satisfies

$$
p(\mathbf{x}_N) = p(\mathbf{y}_N)
$$

when $\sum_{i=1}^{N} x_i = \sum_{i=1}^{N} y_i = N\theta$. Then

$$
p\left(x_1, x_2, \ldots, x_n \mid x_i \geq 0, \sum_{i=1}^{N} x_i = N\theta\right)
$$
$$
= \frac{(N-1)}{N\theta} \frac{(N-2)}{N\theta} \cdots \frac{(N-n)}{N\theta} \left[1 - \frac{\sum_{i=1}^{n} x_i}{N\theta}\right]^{N-n-1}
$$

for $1 \leq n < N$ and $0 \leq \sum_{i=1}^{N} x_i \leq N\theta$.

---

[1]This is not a rigorous proof.

Proof. Since $p(x_1, x_2, \ldots, x_N)$ is constant when $\sum_{i=1}^{N} x_i = N\theta$, we must also have

$$p\left(x_1, x_2, \ldots, x_{N-1} \mid x_i \geq 0, \sum_{i=1}^{N} x_i = N\theta\right) = \text{constant}$$

since $x_N = N\theta - \sum_{i=1}^{N-1} x_i$ when $\sum_{i=1}^{N} x_i = N\theta$.

It will be convenient to let $y_i = \frac{x_i}{N\theta}$ and to compute

$$p\left(y_1, y_2, \ldots, y_{N-1} \mid y_i \geq 0, \sum_{i=1}^{N} y_i = 1\right) \qquad (1.1.3)$$

for $0 \leq \sum_{i=1}^{N-1} y_i \leq 1$. Since (1.1.3) is constant, to compute (1.1.3) we only need to compute the Dirichlet integral

$$\int_{y_1=0}^{1} \int_{y_2=0}^{1-y_1} \cdots \int_{y_{N-1}=0}^{1-y_1-\cdots-y_{N-2}} dy_{N-1} \ldots dy_1$$

$$= \int_{y_1=0}^{1} \int_{y_2=0}^{1-y_1} \cdots \int_{y_{N-2}=0}^{1-y_1-\cdots-y_{N-3}} (1 - y_1 - \cdots - y_{N-2}) dy_{N-2} \ldots dy_1$$

$$= \int_{y_1=0}^{y_1=1} \frac{(1 - y_1)^{N-2}}{(N-2)!} dy_1 = -\frac{(1-y_1)^{N-1}}{(N-1)!} \bigg|_{y_1=0}^{y_1=1} = \frac{1}{(N-1)!} = \frac{1}{\Gamma(N)}.$$

It follows that $p(y_1, y_2, \ldots, y_{N-1} \mid y_i \geq 0, \sum_{i=1}^{N} y_i = 1) = \Gamma(N)$ for $0 \leq \sum_{i=1}^{N-1} y_i \leq 1$.

Integrating in a similar way we can show that

$$p\left(y_1, y_2, \ldots, y_n \mid y_i \geq 0, \sum_{i=1}^{N} y_i = 1\right) = \frac{\Gamma(N)}{\Gamma(N-n)} \left[1 - \sum_{i=1}^{n} y_i\right]^{N-n-1}$$

for $0 \leq \sum_{i=1}^{n} y_i \leq 1$.

Now consider $p(y_1, y_2, \ldots, y_n \mid y_i \geq 0, \sum_{i=1}^{N} y_i = 1) dy_1 \ldots dy_n$ and make the change of variable

$$y_i = \frac{x_i}{N\theta}$$

so that

$$p\left(x_1, x_2, \ldots, x_n \mid x_i \geq 0, \sum_{i=1}^{N} x_i = N\theta\right)$$

becomes

$$= \frac{(N-1)}{N\theta} \cdots \frac{(N-n)}{N\theta} \left[1 - \frac{\sum_{i=1}^{n} x_i}{N\theta}\right]^{N-n-1}. \qquad \square$$

Letting $n = N - 1$, we see that (1.1.1) implies

$$p\left(x_1, x_2, \ldots, x_{N-1} \mid \sum_{i=1}^{N} x_i = N\theta\right) = \frac{\Gamma(N)}{[N\theta]^{N-1}}$$

for $0 \le \sum_{i=1}^{N-1} x_i \le N\theta$. Hence, conditional on θ, the joint conditional density for $\mathbf{x}_{N-1}$ is uniform on the simplex $\{\mathbf{x}_{N-1} \mid x_i \ge 0, \sum_{i=1}^{N-1} x_i \le N\theta\}$.

The above density (1.1.1) is a so-called ℓ_1-isotropic density. The related Dirichlet probability distribution is discussed in DeGroot (1970).

Applications of the finite population exponential model. The strength of structures, such as dikes, deteriorates with time due to environmental factors. Such structures are subject to cumulative deterioration. In the case of dikes, deterioration could be measured by the height of the dike, for example. If the height drops to a specified level, say, h_0, there could be a danger of flooding. Extensive applications of ℓ_1-densities such as (1.1.1) to hydraulic structures are given in the Ph.D. thesis of Jan van Noortwijk (1996).

Let x_i denote the amount of deterioration in the ith unit time interval. The time unit could be months, years, etc. If h denotes the original height, then $h_n = h - \sum_{i=1}^{n} x_i$ denotes the height after n unit time intervals. In this case, h_n depends only on the cumulative deterioration and does not depend on the order in which the deterioration amounts appear. In practice, a lack of deterioration data is common. However, there may be engineering opinions concerning the average rate of deterioration per unit time θ. Using equation (1.1.1), the probability density for

$$s = \sum_{i=1}^{n} x_i$$

can be calculated and used to make decisions concerning inspection intervals and planned maintenance. The time horizon can be specified in terms of N equal time units. The probability density for s is

$$p\left(s \mid x_i \ge 0, \sum_{i=1}^{N} x_i = N\theta\right) = \frac{\Gamma(N)}{\Gamma(n)\Gamma(N-n)} \left[\frac{1}{N\theta}\right]^{n} s^{n-1} \left[1 - \frac{s}{N\theta}\right]^{N-n-1}$$

$$(1.1.4)$$

for $0 \le s \le N\theta$. This is a *beta density* on the interval $[0, N\theta]$ with parameters n and $N - n$, respectively.

The beta density on the interval $[0, 1]$ is

$$p_{A,B}(y) = \frac{\Gamma(A+B)}{\Gamma(A)\Gamma(B)} y^{A-1} [1 - y]^{B-1} \qquad (1.1.5)$$

for $0 \le y \le 1$ and 0 elsewhere with parameters $A, B > 0$. This density will also be useful in later applications.

Exercises

1.1.1. Using computer graphics and (1.1.1), plot $p(x_1 \mid x_i \geq 0, \sum_{i=1}^{N} x_i = N\theta)$ versus $x_1 (0 \leq x_1 \leq N\theta)$ for $N = 2, 5$, and 10 on the same graph. Let $\theta = 10$ in all cases.

1.1.2. Let $X_1, X_2, \ldots, X_n$ be random quantities having joint probability density conditional on $\sum_{i=1}^{N} X_i = N\theta$ given by (1.1.1).

(a) Show that X_1 and X_2 have the same univariate density.

(b) Calculate the conditional mean and variance of the random quantity X_1; i.e., $E(X_1 \mid \sum_{i=1}^{N} X_i = N\theta)$ and $Var(X_1 \mid \sum_{i=1}^{N} X_i = N\theta)$.

1.1.3. Using the definition of Euler's "e" prove (1.1.2).

1.1.4*. Fix $n(1 \leq n < N)$ and let $s = \sum_{i=1}^{n} x_i$. Use the joint density given by (1.1.1) to derive the density $p(s \mid x_i \geq 0, \sum_{i=1}^{N} x_i = N\theta)$ for $0 \leq s \leq N\theta$. Show that

$$\lim_{N \to \infty} p\left(s \mid x_i \geq 0, \sum_{i=1}^{N} x_i = N\theta\right) = \frac{s^{n-1}}{\Gamma(n)\theta^n} e^{-\frac{s}{\theta}}.$$

This is the gamma density with parameters n and $\frac{1}{\theta}$.

1.2. Likelihood.

The conditional density

$$p\left(x_1, x_2, \ldots, x_n \mid x_i \geq 0, \sum_{i=1}^{N} x_i = N\theta\right)$$

$$= \frac{(N-1)}{N\theta} \frac{(N-2)}{N\theta} \cdots \frac{(N-n)}{N\theta} \left[1 - \frac{\sum_{i=1}^{n} x_i}{N\theta}\right]^{N-n-1}$$

for $0 \leq \sum_{i=1}^{n} x_i \leq N\theta$ evaluated at the n observed lifetimes

$$(x_1, x_2, \ldots, x_n)$$

and considered as a function of θ is called the *likelihood* for θ. It is conventional to write

$$L(\theta) = \begin{cases} p(x_1, x_2, \ldots, x_n \mid \sum_{i=1}^{N} x_i = N\theta), & \theta \geq \frac{\sum_{i=1}^{n} x_i}{N}, \\ 0 & \text{otherwise} \end{cases} \qquad (1.2.1)$$

for this likelihood. As a function of θ, it is *not* a probability density in θ. Also it is not, in this case, the product of univariate densities; i.e., the n lifetimes are not conditionally independent given θ.

The reason for our interest in the likelihood is because of its role in Bayes' formula. Let $p(\theta)$ be our initial opinion for θ. Recall that, by Bayes' formula, the posterior density for θ can be calculated as

$$p(\theta \mid x_1, x_2, \ldots, x_n) = \frac{p(x_1, x_2, \ldots, x_n \mid x_i \geq 0, \sum_{i=1}^{N} x_i = N\theta) p(\theta)}{\int_{\theta=0}^{\infty} p(x_1, x_2, \ldots, x_n \mid x_i \geq 0, \sum_{i=1}^{N} x_i = N\theta) p(\theta) d\theta}$$

(1.2.2)

or

$$p(\theta \mid x_1, x_2, \ldots, x_n) \propto L(\theta) p(\theta),$$

where $\propto$ indicates "proportional to," meaning that the right-hand side is missing a factor, not dependent on θ, required to make it a probability density. Since the denominator in equation (1.2.2) does not depend on θ, it can be ignored for most applications where data have been observed. However, in the case of experimental designs, where data have not yet been observed, we would be interested in the exact expression with the denominator intact.

Large population approximation. For large populations, the finite population exponential model (1.1.1) is approximately the infinite population exponential model. In Chapter 2 we refer to the infinite population model as the "exponential model." From the previous section and equation (1.1.2),

$$\lim_{N \uparrow \infty} p \left(x_1, x_2, \ldots, x_n \mid \sum_{i=1}^{N} x_i = N\theta \right) = \prod_{i=1}^{n} \left[\frac{1}{\theta} e^{-\frac{x_i}{\theta}} \right].$$

In Exercise 1.1.1 the reader was asked to graph

$$p \left(x_1 \mid \sum_{i=1}^{N} x_i = N\theta \right) = \frac{N-1}{N\theta} \left[1 - \frac{x_1}{N\theta} \right]^{N-2}$$

for $N = 2, 5$, and 10 when $\theta = 10$. Note that when $N = 2$, the conditional density is the uniform density on $[0, 20]$, while for $N = 10$ the density is already approaching the limiting exponential form although it is only positive for the interval $[0, 100]$.

The likelihood for θ in the limiting case $(N \uparrow \infty)$ is

$$L(\theta) = \prod_{i=1}^{n} p(x_i \mid \theta) = \left(\frac{1}{\theta} \right)^n e^{-\sum_{i=1}^{n} x_i/\theta}.$$

(1.2.3)

In the limiting form, the n lifetimes, before they are observed, are conditionally independent given θ. This is *not true* in the finite population case.

Maximum likelihood. In analyzing data relative to the parameter of interest in this case, namely, θ, it is convenient to find an anchor or initial estimate. The maximum value of $L(\theta)$ provides such a useful location point. In the finite population case,

$$\hat{\theta} = \left(1 - \frac{1}{N}\right) \frac{\sum_{i=1}^{n} x_i}{n}, \tag{1.2.4}$$

while in the limit as $N \uparrow \infty$, $\max_\theta L(\theta)$ is achieved for θ at

$$\hat{\theta} = \frac{\sum_{i=1}^{n} x_i}{n}. \tag{1.2.5}$$

$\hat{\theta}$ is called the maximum likelihood estimator (MLE) for θ and is the mode for the posterior density

$$p(\theta \mid x_1, x_2, \ldots, x_n) \propto p\left(x_1, x_2, \ldots, x_n \mid \sum_{i=1}^{N} x_i = N\theta\right) p(\theta)$$

when $p(\theta)$ is constant over a sufficiently large interval for θ.

Exercises

1.2.1. Show that the MLE for θ in the *finite population* case is $\hat{\theta} = \left(1 - \frac{1}{N}\right) \frac{\sum_{i=1}^{n} x_i}{n}$.

1.2.2. Suppose there are only $N = 10$ special switches produced for a particular space probe. Also suppose $n = 2$ have been tested and have failed after $x_1 = 3.5$ weeks and $x_2 = 4.2$ weeks, respectively.

(a) What is $\hat{\theta}$ in this case?

(b) Graph the likelihood as a function of θ when $\theta = \frac{\sum_{i=1}^{N} x_i}{N}$.

(c) Using a flat prior for θ (i.e., $p(\theta) = $ constant), graph the posterior density for θ given the data.

1.3. Total time on test (TTT) for the finite population exponential model.

Failures and survivors. Often failure data will consist of both item lifetimes and survival times for items that have not yet failed. Since survival times also constitute information, we would like to make use of them. To do this we need to calculate the appropriate likelihood function. This can be done using the following result letting $X_1, X_2, \ldots, X_N$ be the unknown (random) lifetime quantities corresponding to N labelled items.

THEOREM 1.3.1. *Let N be the population size. Fix $\theta > 0$ and assume that we are indifferent to points on the manifold*

$$\left\{\mathbf{x}_N \mid x_i \geq 0, \sum_{i=1}^{N} x_i = N\theta\right\}$$

in the sense that, conditional on θ, our joint probability function $p(\bullet)$ satisfies

$$p(\mathbf{x}_N) = p(\mathbf{y}_N)$$

when $\sum_{i=1}^{N} x_i = \sum_{i=1}^{N} y_i = N\theta.$ *Then*

$$P\left(X_1 > y_1, X_2 > y_2, \ldots, X_n > y_n \mid \sum_{i=1}^{N} X_i = N\theta\right)$$

$$\stackrel{DEF}{=} \overline{F}(y_1, y_2, \ldots, y_n \mid \theta) = \left[1 - \sum_{i=1}^{n} \frac{y_i}{N\theta}\right]^{N-1}$$

(1.3.1)

for $y_i \geq 0, 0 \leq \sum_{i=1}^{n} y_i \leq N\theta.$ *(The expression (1.3.1) is also valid when* $n = N.$*)*

Proof. Since the y_i's are survival ages and *not* failure ages (i.e., the previous x_i's) we have used a different notation. To show (1.3.1), take the n-fold partial derivative of $[1 - \sum_{i=1}^{n} \frac{y_i}{N\theta}]^{N-1}$ and, multiplying by $(-1)^n$ since $[1 - \sum_{i=1}^{n} \frac{y_i}{N\theta}]^{N-1}$ is claimed to be the joint survival probability, we have

$$(-1)^n \frac{\partial^n}{\partial y_1 \partial y_2 \ldots \partial y_n} \left[1 - \sum_{i=1}^{n} \frac{y_i}{N\theta}\right]^{N-1} \Bigg|_{y_1 = x_1, \ldots, y_n = x_n}$$

$$= \frac{(N-1)}{N\theta} \cdots \frac{(N-n)}{N\theta} \left[1 - \frac{\sum_{i=1}^{n} x_i}{N\theta}\right]^{N-n-1}$$

$$= p\left(x_1, x_2, \ldots, x_n \mid x_i \geq 0, \sum_{i=1}^{N} x_i = N\theta\right).$$

Evaluating the partial derivative result at $y_1 = x_1, y_2 = x_2, \ldots, y_n = x_n$ we have the joint conditional density given in Theorem 1.1.1. This proves (1.3.1). $\square$

Again, suppose n devices are under observation with respect to lifetime, but now we are required to analyze the data obtained before all devices have failed. Suppose

$$x_{(1)} \leq x_{(2)} \leq \cdots \leq x_{(r)}$$

are the first r observed failure ages while $n - r$ devices are still operating at age $x_{(r)}$. (These are the first r *order statistics*.)

We require the likelihood for θ with respect to the finite population exponential model $p(x_1, x_2, \ldots, x_n \mid \sum_{i=1}^{N} x_i = N\theta)$. In this case it is unnecessary to calculate the exact conditional probability for the event: r items fail at ages $x_{(1)} \leq x_{(2)} \leq \cdots \leq x_{(r)}$ and $n - r$ survive to age $x_{(r)}$ given $\sum_{i=1}^{N} x_i = N\theta$. It is enough to assume that the first r labelled devices failed at ages $x_1, x_2, \ldots, x_r$ and the remaining $n - r$ devices survived to age x_r and calculate the conditional probability of this event. To obtain the exact conditional probability with respect to $x_{(1)} \leq x_{(2)} \leq \cdots \leq x_{(r)}$ we need only multiply this calculation by $\frac{n!}{1!1!\ldots1!(n-r)!}$, where there are r terms 1! in the denominator. However, since this factor does not involve θ, it will cancel when we compute the posterior.

Using (1.3.1) we take the r-fold partial derivative with respect to $y_1, y_2, \ldots, y_r$, letting $y_1 = x_1, y_2 = x_2, \ldots, y_r = x_r, y_{r+1} = y_{r+2} \cdots = y_n = x_r$

and multiplying by $(-1)^r$ to obtain

$$\frac{(N-1)}{N\theta}\frac{(N-2)}{N\theta}\cdots\frac{(N-r)}{N\theta}\left[1 - \frac{\sum_{i=1}^{r} x_i + \sum_{i=r+1}^{n} x_r}{N\theta}\right]^{N-r-1}.$$

The conditional joint probability density for the event $x_{(1)} \leq x_{(2)} \leq \cdots \leq x_{(r)}$ and $n-r$ survive to age $x_{(r)}$ given $\sum_{i=1}^{N} x_i = N\theta$ is now proportional to

$$\left(\frac{1}{\theta}\right)^r \left(1 - \frac{\sum_{i=1}^{r} x_{(i)} + (n-r)x_{(r)}}{N\theta}\right)^{N-r-1} dx_{(1)}\ldots dx_{(r)}, \qquad (1.3.2)$$

where we have included $dx_{(1)}dx_{(2)}\ldots dx_{(r)}$ since (1.3.2) is actually a mixed probability density and a survival probability, namely the probability that r items individually fail in the intervals

$$[x_{(1)}, x_{(1)} + dx_{(1)}), [x_{(2)} + dx_{(2)}), \ldots, [x_{(r)} + dx_{(r)})$$

and that $n-r$ items survive to age $x_{(r)}$.

The likelihood $L(\theta)$ satisfies

$$L(\theta) \propto \left(\frac{1}{\theta}\right)^r \left(1 - \frac{\sum_{i=1}^{r} x_{(i)} + (n-r)x_{(r)}}{N\theta}\right)^{N-r-1}. \qquad (1.3.3)$$

The MLE for θ in this case is

$$\hat{\theta} = \left(1 - \frac{1}{N}\right)\frac{T(x_{(r)})}{r}, \qquad (1.3.4)$$

where $T(x_{(r)}) = \sum_{j=1}^{r} x_{(j)} + (n-r)x_{(r)}$ is the total observed lifetime of all n devices under observation. We call $T(x_{(r)})$ the TTT statistic, so that $\hat{\theta}$ is the TTT divided by the number of observed failures modified by the factor $(1 - \frac{1}{N})$. TTT is merely our name for this quantity since not all items need be put on life test at the same time.

The *sufficient statistic* for θ in this case is

$$\left(r, T(x_{(r)})\right),$$

the number r of observed failures, and the TTT $T(x_{(r)})$. The statistic is sufficient since these are the only quantities, based on the data, needed to evaluate the likelihood.

Exercises

1.3.1. Derive the MLE

$$\hat{\theta} = \left(1 - \frac{1}{N}\right)\frac{T(x_{(r)})}{r}$$

in equation (1.3.4).

1.3.2. Graph

$$r\left(x_1 \mid \sum_{i=1}^{N} x_i = N\theta\right) = \frac{p(x_1 \mid \sum_{i=1}^{N} x_i = N\theta)}{\overline{F}(x_1 \mid \theta)}$$

for $0 \le x_1 < N\theta$ when $N = 5$ and $\theta = 10$. We call $r(x_1 \mid \sum_{i=1}^{N} x_i = N\theta)$ the "failure rate" function for x_1. In the finite population case, it is of no significance although it is considered useful in the infinite population case.

1.3.3. For the finite population case and using

$$\overline{F}(y_1, y_2, \dots, y_n \mid \theta) = \left[1 - \sum_{i=1}^{n} \frac{y_i}{N\theta}\right]^{N-1}$$

for $y_i \ge 0, 0 \le \sum_{i=1}^{n} y_i \le N\theta$, perform the following calculations.

(a) Calculate the likelihood function for θ when the data consists of r observed failure ages and $n-r$ observed survival ages. Let $x_1, x_2, \dots, x_r$ denote the failure ages and $y_{r+1}, y_{r+2}, \dots, y_n$ the survival ages. Survival ages can be less than, equal to, or greater than failure ages.

(b) Show that the MLE for θ in this case is

$$\hat{\theta} = \left(1 - \frac{1}{N}\right)\frac{T}{r},$$

where $T = \sum_{i=1}^{r} x_i + \sum_{i=r+1}^{n} y_i$.

1.3.4. Using (1.1.1) calculate the exact joint probability density for the order statistics $x_{(1)} \le x_{(2)} \le \dots \le x_{(n)}$ from a sample of size n given a population of size N and conditional on $\sum_{i=1}^{N} x_i = N\theta$.

1.3.5*. Show that (1.3.1) still holds when $n = N$; i.e., show that

$$P\left(X_1 > y_1, X_2 > y_2, \dots, X_N > y_N \mid \sum_{i=1}^{N} X_i = N\theta\right)$$

$$= \left[1 - \sum_{i=1}^{N} \frac{y_i}{N\theta}\right]^{N-1}$$

for $y_i \ge 0, 0 \le \sum_{i=1}^{N} y_i \le N\theta$.

1.4. Exchangeability.

In Theorem 1.3.1, we derived the joint density of random quantities under the indifference judgment relative to sums. Obviously in calculating a sum, the order in which we do the sum is immaterial. The joint density in this case is said to be *exchangeable*.

Let N items or devices be labelled $\{1, 2, \dots, N\}$. Let $X_1, X_2, \dots, X_N$ be potential measurements on these items with respect to some quantity of interest. Then the *items* are said to be *exchangeable with respect to this quantity of interest* if in our judgment

$$P(X_1 > x_1, X_2 > x_2, \ldots, X_N > x_N) = P(X_{\pi_1} > x_1, X_{\pi_2} > x_2, \ldots, X_{\pi_N} > x_N) \tag{1.4.1}$$

for our generic joint survival probability function $P(\bullet)$ for any permutation $(\pi_1, \pi_2, \ldots, \pi_N)$ of the integers $\{1, 2, \ldots, N\}$ and all vectors $\mathbf{x}_N$. In other words, in our judgment items are similar, and so order does not count in the joint probability function.

Notice that it is the items that are actually exchangeable and only with respect to a specified quantity of interest such as lifetime. If we were, for example, interested in lifetimes, the colors of the items might not be relevant so far as exchangeability with respect to lifetime is concerned. Also, the measurements are potential. Once the measurements are known, the items are no longer necessarily exchangeable. Although exchangeability is a probabilistic expression of a certain degree of ignorance, it is a rather strong judgment and in the case of 0-1 random quantities implies a joint probability function related to the hypergeometric distribution as we show in Chapter 3, section 3.1.

The judgment of exchangeability implies that univariate, bivariate, etc. distributions are all the same. It does not imply that exchangeable random quantities are necessarily independent. However, iid random quantities are exchangeable.

Randomness. Randomness, like probability, of which it is a special case, is an expression of a relationship between the analyst and the world. It is not a property of the world, which is why mathematicians have had such trouble defining it.

A sequence of 0's and 1's is random for the analyst if, given the values of the sequence at some places, the analyst would think another place is equally likely to be occupied by a 0 or a 1. In this case, the random quantities, the X_i's, are the 0's and 1's in a sequence. The random quantities are also exchangeable, but the concept is stronger since it implies independence and specific marginal probabilities, namely,

$$P(X_i = 1) = P(X_i = 0) = \frac{1}{2}.$$

By a *random sample* of lifetimes, we mean a sequence $\{X_1, X_2, \ldots, X_n\}$ that is *exchangeable*. Marginal distributions are unknown, and therefore the random quantities are dependent. Independence is impossible since knowledge of some lifetimes would imply partial knowledge of the univariate marginal distribution which would influence our prediction of future lifetimes.

Exercises

1.4.1. In the example concerning dike deterioration in section 1.1, what are the exchangeable "items," and with respect to what are they exchangeable?

1.4.2 (finite exchangeability). Let $P[X_1 = 1, X_2 = 0] = P[X_1 = 0, X_2 = 1] = \frac{1}{2}$. Show that X_1 and X_2 are negatively correlated and exchangeable. That is show that $Cov(X_1, X_2) = E[(X_1 - E(X_1))(X_2 - E(X_2))] < 0$.

1.4.3* (infinite exchangeability). Let $X_1, X_2, \ldots$ be an infinite sequence of binary (0 or 1) exchangeable random quantities. If binary (0 or 1) random quantities or events are judged exchangeable and there are a conceptually infinite number of them, then the probability that k-out-of-n are 1 is necessarily of the form

$$P\left(\sum_{i=1}^{n} X_i = k\right) = \binom{n}{k} \int_0^1 \pi^k (1 - \pi)^{n-k} p(\pi) d\pi \qquad (1.4.2)$$

for some density $p(\bullet)$ (or, more generally, a distribution). Bayesians would interpret $p(\bullet)$ as a prior for π. This was proved by Bruno de Finetti (1937).

Show that for infinite exchangeability $Cov(X_i, X_j) \geq 0$ for any i, j pair ($i \neq j$) in the sequence. $Cov(\bullet, \bullet)$ means covariance of the quantities in question. Hence random quantities in an infinite exchangeable sequence are positively correlated unless the sequence is iid, in which case $Cov(X_i, X_j) = 0$ ($i \neq j$).

Hint. You may want to use the following formula with respect to three related random quantities, $X, Y,$ and Z:

$$Cov(X, Y) = E_Z Cov(X, Y \mid Z) + Cov_Z[E(X \mid Z), E(Y \mid Z)].$$

1.4.4. Let $X_1, X_2, \ldots$ be an infinite sequence of binary (0 or 1) exchangeable random quantities as in Exercise 1.4.3*. Suppose we observe $x_1, x_2, \ldots, x_n$ and wish to predict $x_{n+1}, x_{n+2}, \ldots, x_N$. Let $\sum_{i=1}^{n} x_i = k$ and $\sum_{i=1}^{N} x_i = k + m$.

Using (1.4.2) and the definition of conditional probability we have

$$p(x_{n+1}, x_{n+2}, \ldots, x_N \mid x_1, x_2, \ldots, x_n) = \frac{\binom{N}{m+k} \int_0^1 \pi^{m+k}(1-\pi)^{N-m-k} p(\pi) d\pi}{\binom{n}{k} \int_0^1 \pi^k (1-\pi)^{n-k} p(\pi) d\pi}.$$

Show that this implies

$$p(x_{n+1}, x_{n+2}, \ldots, x_N \mid x_1, x_2, \ldots x_n)$$
$$= \int_0^1 \binom{N-n}{m} \pi^m (1 - \pi)^{N-n-m} p(\pi \mid x_1, x_2, \ldots, x_n) d\pi,$$

where $p(\pi \mid x_1, x_2, \ldots, x_n)$ can be obtained using Bayes' formula (1.2.2).

1.5.   Notes and references.

Section 1.1. The main reason for starting with the finite population exponential model rather than with the infinite population exponential model is to better understand the judgmental basis for probability distribution models.

Many of the ideas in this section and the chapter are due to M. B. Mendel. His 1989 Ph.D. thesis, *Development of Bayesian Parametric Theory with Applications to Control*, contains a generalization of Theorem 1.1.1. J. M. van Noortwijk in his 1996 Ph.D. thesis, *Optimal Maintenance Decisions for Hydraulic Structures under Isotropic Deterioration* applied many of the ideas of Mendel's thesis to maintenance problems involving dikes. Generic joint probability densities are ℓ_q isotropic if

$$p(x_1, x_2, \ldots, x_N) = p(y_1, y_2, \ldots, y_N)$$

when $\sum_{i=1}^{N} |x_i|^q = \sum_{i=1}^{N} |y_i|^q$. When $q = 1$ and $x_i \geq 0$, $i = 1, 2, \ldots, N$ we have the finite population exponential probability distribution.

Section 1.2. The idea of deriving the likelihood from indifference principles was the main theme in Mendel's 1989 thesis. This approach has been called the *operational Bayesian approach.*

Section 1.3. In the case of the infinite population exponential model $p(x \mid \theta) = \frac{1}{\theta} e^{-x/\theta}$, the sufficient statistics for the parameter θ are also the number of observed failures together with the "TTT."

Section 1.4. The concept of exchangeability, due to B. de Finetti (1937), replaces the usual iid assumption in many respects. It is used throughout Part I of this book. The discussion of randomness is due to Dennis Lindley. The importance of exchangeability in inference was discussed in a 1981 paper by Lindley and Novick, "The Role of Exchangeability in Inference."

Notation

$x_1, x_2, \ldots, x_N$	vector elements or arguments (not random quantities)
$x_{(1)} \leq x_{(2)} \leq \cdots \leq x_{(N)}$	order statistics based on $x_1, x_2, \ldots, x_N$ (i.e., ordered observations)
$\mathbf{x}_N$	boldface denotes a vector; the vector has size N
$X_1, X_2, \ldots, X_N$	random quantities (also called random variables)
$P(E)$	probability of event E
$p(x \mid \theta)$	the conditional probability of the random quantity X given the random quantity θ; this is an abuse of notation wherein the arguments tell us which random quantities are being considered
$p(x_1, x_2, \ldots x_n)$	a joint probability density function
θ, λ	parameters, usually θ = population mean and $\lambda = \frac{1}{\theta}$
$\hat{\theta}$	MLE of θ

$p(x_1, x_2, \ldots, x_n \mid x_i \geq 0, \sum_{i=1}^{N} x_i = N\theta)$	a joint conditional probability density
$p(\theta \mid x_1, \ldots, x_n)$	posterior density for θ
$p(x_{n+1} \mid x_1, \ldots, x_n)$	predictive density
E	expectation, i.e., if X has density $p(\bullet)$, $E(X) = \int_0^\infty x p(x) dx$
$L, L(\theta)$	likelihood, likelihood of θ
$T, T(x_{(r)})$	TTT, TTT until $x_{(r)}$
iid	independent, identically distributed refers to random quantities
$Cov(X, Y)$	covariance of random quantities X, Y
$p(x \mid \theta) = \frac{1}{\theta} e^{-x/\theta}$	infinite population exponential model
$p_{A,B}(y) = \dfrac{\Gamma(A+B)}{\Gamma(A)\Gamma(B)} y^{A-1}(1-y)^{B-1}$	beta density for $0 \leq y \leq 1; A, B > 0$
	mean $= \dfrac{A}{A+B}$, variance $= \dfrac{AB}{(A+B)^2(A+B+1)}$

CHAPTER **2**

Lifetime Data Analysis

In this chapter we assume that populations are effectively unbounded; i.e., we are assuming the limiting population case ($N \uparrow \infty$).

2.1. Data analysis for the exponential model.

In the previous chapter we *derived* the finite population exponential model for populations of size N. Only in the limit, as $N \uparrow \infty$, did we obtain the usual exponential model, i.e., in the one-dimensional case where θ is the limiting average lifetime

$$p(x \mid \theta) = \frac{1}{\theta} e^{-\frac{x}{\theta}} \tag{2.1.1}$$

for $x \geq 0$ and $\theta > 0$. We call (2.1.1) the "exponential model."

The first idea in Chapter 1 was that the derivation should be based on a judgment of indifference on submanifolds. The importance of this derivation was due to the fact that we focused on the average lifetime of the population. Were we to focus on other aspects of the lifetimes of the population, we would have derived a different conditional probability model.

The second idea was that when we insert data in our conditional probability model, the corresponding function of θ, called the likelihood $L(\theta)$, provides the key tool in analyzing data. (In Chapter 1, θ was the population lifetime average.) Using the likelihood and a prior opinion concerning θ we can compute the posterior probability function for θ and from this answer any probability question we may have concerning θ or future lifetimes.

The influence of failures on the posterior density. Suppose NO failures are observed but all items have survived for some time t. How can we analyze item lifetimes in this case? To make the situation clear we will consider a particular case. We will show that, although survivors tend to increase our estimate for the mean life θ, lack of failures also increases our uncertainty concerning θ. While the engineer naturally wants to see no failures, the statistician–analyst wants to see failures in order to decrease the uncertainty in the analysis.

Example 2.1.1. Failures and survivors and the exponential model.
Suppose we put $n = 10$ devices on life test. Suppose furthermore that the first failure occurs at age $x_{(1)} = 1$ and a second failure occurs at $x_{(2)} = 3$. Let observation cease at time t. What can we say when $t < 1$? What if $1 < t < 3$? How does our inferential opinion about θ change as we go further into the life test?

In general, as we shall see, observed failure times sharpen our posterior density for θ while survivors tend to move the mode of the posterior density to the right but also result in a more *diffuse* probability function.

To understand this phenomenon, we will consider the infinite population approximation, namely, the exponential model

$$p(x \mid \theta) = \frac{1}{\theta} e^{-\frac{x}{\theta}}.$$

To illustrate, we will use the so-called natural conjugate prior density for θ, namely,

$$\pi_{a,b}(\theta) = \frac{b^a \theta^{-(a+1)} e^{-\frac{b}{\theta}}}{\Gamma(a)} \tag{2.1.2}$$

for $a, b > 0$. This is called the inverted gamma density, since, if θ is a random quantity with density $\pi_{a,b}(\theta)$, then

$$\theta^{-1} = \lambda$$

has a gamma density. We call λ the failure rate. It only applies to the infinite population model. In this case,

$$r(x \mid \theta) = \frac{p(x \mid \theta)}{P(X > x \mid \theta)} = \frac{1}{\theta} = \lambda$$

is the failure rate function. Note that $r(x \mid \theta) = \frac{1}{\theta}$ is constant in x.

For a random sample of n lifetimes, let

$$0 \equiv x_{(0)} \leq x_{(1)} \leq x_{(2)} \leq \cdots \leq x_{(r)} < t,$$

where only r of n possible lifetimes have been observed while $n - r$ items survive the interval $(0, t]$. Observation stops at time t. The likelihood $L(\theta)$ is proportional to

$$\left[\prod_{i=1}^{r} e^{-\frac{x_{(i)}}{\theta}} \right] e^{-\frac{(n-r)t}{\theta}} = \frac{1}{\theta^r} e^{-\frac{T(t)}{\theta}}, \tag{2.1.3}$$

where $T(t) = \sum_{i=1}^{r} x_{(i)} + (n - r)t$ is the total time on test (TTT) to age t.

The sufficient statistic for θ is $(r, T(t) = T)$. If $\pi_{a,b}(\theta)$ is the inverted gamma prior for θ, then the posterior for θ is again the inverted gamma but now with parameters

$$a + r \quad \text{and} \quad b + T.$$

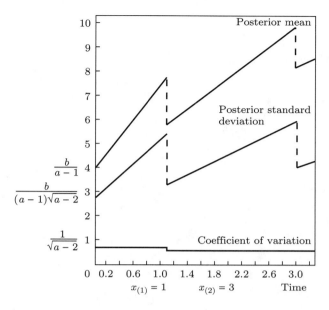

FIG. 2.1.1.   *Posterior mean, posterior standard deviation, and posterior coefficient of variation as a function of elapsed test time.*

The mean of the inverted gamma density is readily calculated:

$$\int_0^\infty \theta \pi_{a,b}(\theta)\, d\theta = \frac{b}{a-1}.$$

For the inverted gamma posterior density in which the parameters are $a+r$ (in place of a) and $b+T$ (in place of b), the corresponding mean takes the form

$$\frac{b+T}{a+r-1} \equiv (1-w)\frac{b}{a-1} + w\frac{T}{r},$$

where $w = \frac{r}{a+r-1}$. Thus, the mean of the posterior density may be written as a convex combination of the prior mean $\frac{b}{a-1}$ and the maximum likelihood estimate (MLE) $\frac{T}{r}$ of the exponential life distribution mean parameter. Note that, as r, the number of observed failures increases; the posterior mean attaches more weight to the MLE and less weight to the prior mean.

Example 2.1.2.   In order to provide an explicit answer to the questions posed in Example 2.1.1, let θ have the inverted gamma prior with parameters $a = 4$ and $b = 12$. In Figure 2.1.1 we plot the posterior mean, posterior standard deviation, and posterior coefficient of variation as a function of t, the test time elapsed.

Note that as we go into the test both the posterior mean and the posterior standard deviation increase with test time t. However, as soon as a failure

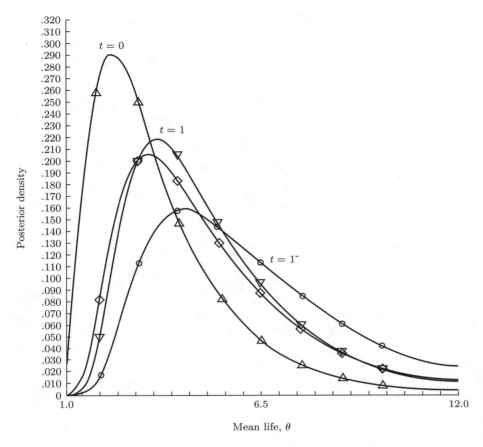

FIG. 2.1.2. *Posterior densities at selected test times (t).*

occurs, the posterior mean, the posterior standard deviation, and the coefficient of variation decrease.

In Figure 2.1.2 we have plotted the posterior density for θ at selected times during the life test. The posterior density for $t = 0$ is, of course, the prior density. Notice the shape of the posterior density at $t = 1^-$ (i.e., just before the first observed failure) and at $t = 1$ (i.e., just after the first failure).

Table 2.1.1 summarizes the properties of the natural conjugate prior density and of the corresponding posterior density for the two possible parametrizations of the exponential model.

The predictive density for the exponential model. In the infinite population case, the univariate marginal density is $p(x \mid \theta) = \frac{1}{\theta} e^{-x/\theta}$. If the prior for θ, $p(\theta)$, is the inverse gamma with parameters a and b (i.e., $IVG(a,b)$), then $p(\theta \mid r, T)$ is $IVG(a + r, b + T)$, where r is the number of observed failures and T is the TTT. The predictive density

TABLE 2.1.1

Parameter	Failure rate λ	Mean life θ
Likelihood	$\lambda^r e^{-\lambda T}$	$\theta^{-r} e^{-\frac{T}{\theta}}$
Natural conjugate	$\dfrac{b^a \lambda^{a-1} e^{-b\lambda}}{\Gamma(a)}$	$\dfrac{b^a \theta^{-(a+1)} e^{-\frac{b}{\theta}}}{\Gamma(a)}$
Prior	(gamma)	(inverted gamma)
Prior mean	$\dfrac{a}{b}$	$\dfrac{b}{a-1}, a > 1$
Prior variance	$\dfrac{a}{b^2}$	$\dfrac{b^2}{(a-1)^2(a-2)}, a > 2$
Prior coefficient of variation	$a^{-\frac{1}{2}}$	$(a-2)^{-\frac{1}{2}}, a > 2$
Posterior mean	$\dfrac{(a+r)}{(b+T)^2}$	$\dfrac{(b+T)}{(a+r-1)}$
Posterior variance	$\dfrac{a+r}{(b+T)^2}$	$\dfrac{(b+T)^2}{(a+r-1)^2(a+r-2)}$
Posterior coefficient of variation	$(a+r)^{-\frac{1}{2}}$	$(a+r-2)^{-\frac{1}{2}}$

Note. r = number of observed failures. T = TTT.

$$p(x) = \int_0^\infty p(x \mid \theta) p(\theta) \, d\theta$$

is *Pareto*(a, b); i.e.,

$$p(x) = \frac{\Gamma(a+1)}{\Gamma(a)} \frac{b^a}{(x+b)^{a+1}}$$

for $x \geq 0$. Having observed x_1, $p(x \mid x_1)$ is *Pareto*$(a+1, b+x_1)$ so that $p(x \mid x_1)$ is different from $p(x)$.

Suppose now that the lifetime X of an item is judged exponential and is observed to survive to time t. Then

$$P(X > x + t \mid X > t, \theta) = \frac{e^{-\frac{(x+t)}{\theta}}}{e^{-\frac{t}{\theta}}} = e^{-\frac{x}{\theta}} = P(X > x \mid \theta) \qquad (2.1.4)$$

so that $X - t$ conditional on survival to time t, conditional on θ, has the same distribution as X conditional on θ. But unconditional on θ we have

$$E(X - t \mid X > t) = E_\theta[E(X - t \mid X > t, \theta)] = E_\theta[\theta \mid X > t] > E(\theta) = E(X) \qquad (2.1.5)$$

so that X and $X - t$ conditional on survival to time t do *not* have the same distribution. Equation (2.1.5) is true for arbitrary nondegenerate prior distributions for θ [Barlow and Proschan (1985)]. It is obvious in the case of the $IVG(a, b)$ prior for θ.

The result given by (2.1.4) is called the *memoryless property* of the exponential model. It is only valid conditional on θ. It is *not* true for the finite population exponential model derived in Chapter 1.

No planned replacement.   A curious property of the infinite population exponential model is that, under this model judgment, we would *never* plan replacement before failure. We would only replace at failure. This is regardless of the cost of a failure during operation of the item. The result follows from (2.1.4) and (2.1.5).

In the case of the finite population exponential model we would, in general, plan replacement, depending on costs involved. This is intuitively more reasonable and should serve as a warning relative to the indiscriminate use of infinite population approximation models.

Exercises

2.1.1. Show that if θ has the inverted gamma density

$$\pi_{a,b}(\theta) = b^a \theta^{-(a+1)} \exp\left[-\frac{b}{\theta}\right] / \Gamma(a),$$

then $\lambda = \theta^{-1}$ has a gamma density

$$p(\lambda \,|\, a, b) = \frac{b^a \lambda^{a-1} e^{-b\lambda}}{\Gamma(a)}$$

with mean $\frac{a}{b}$ and variance $\frac{a}{b^2}$. (*Hint.* Start with $\pi_{a,b}(\theta)\, d\theta$.)

2.1.2. Calculate the *predictive density* for X, i.e.,

$$p(x) = \int_0^\infty p(x \,|\, \theta)\pi(\theta)\, d\theta$$

when $p(x \,|\, \theta) = \frac{1}{\theta}e^{-x/\theta}$ and

$$\pi(\theta) = b^a \theta^{-(a+1)} \exp\left[-\frac{b}{\theta}\right] / \Gamma(a).$$

This is called the Pareto probability density.

2.1.3. Show that if θ has an inverted gamma prior density, r failures are observed, and T is the "TTT," then $p(\theta \,|\, r, T)$ is an inverted gamma density with parameters $a + r$ and $b + T$.

2.1.4. In Example 2.1.1 with $n = 10$ devices on life test, suppose that an additional failure is observed at age $x_{(3)} = 5$. Graph the posterior density for θ at ages $t = 5^-$ and $t = 5$.

2.1.5. For the exponential model $p(x \mid \theta) = \frac{1}{\theta} e^{-x/\theta}$, suppose we start with an arbitrary prior density, $\pi_o(\theta)$, for θ. Three alternative test plans are to be compared as follows.

(1) Observe n units and cease observation at the r_1th failure. Let the sufficient statistic be (r_1, T_1). Compute the posterior density $\pi_1(\theta \mid r_1, T_1)$. Take a second sample of m units and cease observation at the r_2th failure. Let the sufficient statistic be (r_2, T_2). Using $\pi_1(\theta \mid r_1, T_1)$ as if it were a prior density, compute an updated posterior density $\pi_2(\theta \mid n, r_1, m, r_2)$.

(2) Proceed as in plan (1) but with the computations in the reverse order.

(3) Finally, combine both samples obtaining (n, r_1, m, r_2) with sufficient statistic $(r_1 + r_2, T_1 + T_2)$. Starting with the original prior density $\pi_o(\theta)$, obtain the posterior density for θ using the data (n, r_1, m, r_2) from the pair of tests.

Compare the posterior densities under the three test plans.

2.1.6. Make a graph similar to Figure 2.1.1 including a withdrawal at time $t = 2$ as well as the failures at $t = 1$ and $t = 3$ in Example 2.1.1. A withdrawal does not increase the number of observed failures since it is not a failure. How does this effect the posterior mean, standard deviation, and coefficient of variation?

2.2. TTT plots.

In the introduction we briefly discussed the frequency interpretation of probability. In this interpretation, lifetimes of items are considered to be outcomes of repetitive trials generated by some "probability mechanism." This is similar to the numbers that would be generated by Monte Carlo trials on a computer using a random number generator. This is the interpretation of probability which resulted from mathematical studies of gambling.

In his fundamental paper on probability, "Foresight: Its Logical Laws, Its Subjective Sources" (1937), Bruno de Finetti found the connection between the subjective concept of probability and the frequentist concept of probability. The connection is based on the concept of *exchangeable* random quantities. Random quantities $X_1, X_2, \ldots, X_n$ are said to be exchangeable if and only if their joint probability measure is invariant relative to their order. That is, the *joint* cumulative distribution of lifetimes is invariant under permutations of the arguments; i.e., $F(x_1, x_2, \ldots, x_n) = F(x_{\pi_1}, x_{\pi_2}, \ldots, x_{\pi_n})$. It follows from this judgment that all univariate (and multivariate) marginal distributions are the same. Exchangeability does not imply independence. However, iid random quantities are exchangeable.

In 1937 de Finetti showed that an *infinite* sequence of exchangeable random quantities $X_1, X_2, \ldots, X_n, \ldots$ are independent, conditional on the common univariate marginal distribution F. In introducing the concept of exchangeability, de Finetti proves both the *weak and strong laws of large numbers* under the judgment of exchangeability and finite first and second cross moments of the exchangeable random quantities. These "laws" are valid conditional on the univariate marginal distribution F. This is the mathematical result which "justifies," to a very limited extent, the TTT plotting method for analyzing life data

TABLE 2.2.1
Jet engine corrective maintenance removal times.

Corrective maintenance removal times	
ABC123	389
ABC124	588
ABC133	840
ABC134	853
ABC135	868
ABC137	931
ABC139	997
ABC141	1060
ABC146	1200
ABC147	1258
ABC148	1337
ABC149	1411

which we consider next. It doesn't really work since F is unknown. However it does provide an initial analysis of data without a priori judgment. Such an analysis should always be followed by a finer in-depth bayesian analysis. The following example illustrates the TTT plotting method.

Example 2.2.1. Jet engine life.   The turbofan jet engine entered service over 20 years ago as the modern means of propulsion for commercial aircraft. It has provided a very economical and reliable way of transporting cargo and people throughout the world. Maintenance on such jet engines is determined partly by the operating time or the "time-on-wing" (TOW) measurement. If removal of an engine requires penetration into a module (standardized breakdown of the engine into workable sections), then it is classified as a shop visit. Such corrective maintenance removals will be considered failure times in our analysis. The following data on $n = 12$ corrective maintenance removal times in operating hours was recorded in a recent paper on the subject. See Table 2.2.1.

The question we wish to answer is the following: Were we to observe a very large population of removal times exchangeable with these observations, would the empirical cumulative life distribution based on such a large population look like the exponential cumulative distribution? Obviously a definitive answer based on only 12 observations is not possible. However, there is a plotting technique that can at least give us a clue to the answer.

Scaled TTT plots of data.   A scaled TTT data plot is a graph based on the data that tends to look like the scaled TTT transform of some univariate distribution. This univariate distribution may be suggestive of a suitable conditional distribution leading to a likelihood model. It may be useful in a preliminary data analysis. It is only conditional, however, since the plot is scale invariant; i.e., multiplying all observations by the same constant would not change the plot. Hence we can only "identify" a probability distribution up to a scale parameter. It is important to emphasize that the plotting method depends on the a priori judgment that items are exchangeable with respect to lifetime.

Consider the corrective maintenance removal data for jet engines in Table 2.2.1. In that case $n = 12$. The ordered removal times correspond to the ages (measured in TOW hours) at which the jet engines were removed for corrective maintenance. To plot the sample data, first order the observed failure times in increasing order. Let $0 \equiv x_0 \leq x_{(1)} \leq x_{(2)} \leq \cdots \leq x_{(n)}$ be the n ordered failure ages. Let $n(u)$ be the number of items "under test" at age u, i.e., not in need of corrective removal. Then

$$T(t) \stackrel{DEF}{=} \int_0^t n(u)\, du$$

is the TTT to age t (actually $T(t)$ is piecewise linear since $n(u)$ is a step function). If $x_{(r)} \leq t < x_{(r+1)}$ then

$$T(t) = \sum_{i=1}^r (n - i + 1)(x_{(i)} - x_{(i-1)}) + (n - r)(t - x_{(r)}). \qquad (2.2.1)$$

Plot

$$\frac{T(x_{(i)})}{T(x_{(n)})} \qquad \text{versus} \qquad \frac{i}{n}$$

for $i = 1, 2, \ldots, n$. The plot is a function on $[0, \frac{1}{n}, \frac{2}{n}, \ldots, 1]$ to $[0, 1]$ so that no additional scaling is necessary. Figure 2.2.1 is the scaled TTT plot of the data in Table 2.2.1 where we have joined plotted points by straight line segments.

Scaled TTT plot for exponential data. To get some idea of the meaning of the plot in Figure 2.2.1, suppose we generate random exponentially distributed numbers on a computer. If X is distributed as an exponential random quantity with distribution $G(x \mid \theta = 1) = 1 - e^{-x}$, then $U \stackrel{DEF}{=} G(X \mid \theta = 1)$ is distributed as a uniform random variable on $[0, 1]$. Suppose we use this fact to generate $n = 12$ exponentially distributed random quantities. Such random outcomes are listed in Table 2.2.2.

Figure 2.2.2 is a scaled TTT plot of the data in Table 2.2.2. Were we to generate another random sample from $G(x \mid \theta = 1) = 1 - e^{-x}$, we would construct a plot somewhat different from Figure 2.2.2. However, as we shall show, such exponential plots tend to "wind around" the $45°$ line. In contrast, Figure 2.2.1 shows no such tendency. As we shall argue, a plot of the type shown in Figure 2.2.1 indicates wearout in the sense that older engines are more likely to fail than newer engines.

TTT Plot $n = 12$

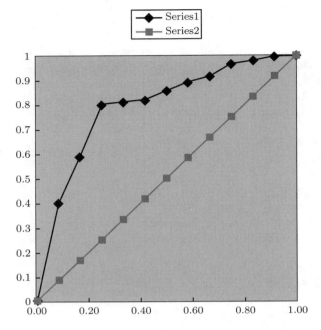

FIG. 2.2.1. *Scaled TTT plot of engine data (Series 1) and exponential transform (Series 2).*

TABLE 2.2.2
Ordered random quantities distributed as $G(x \mid \theta = 1) = 1 - e^{-x}$.

1	0.001
2	0.17
3	0.25
4	0.41
5	0.44
6	0.71
7	0.78
8	0.97
9	1.26
10	1.44
11	1.65
12	1.78

The empirical probability distribution. The empirical distribution is, by construction, a cumulative probability distribution. It is based on the order statistics

$$0 \equiv x_{(0)} \leq x_{(1)} \leq x_{(2)} \leq \cdots \leq x_{(n)}$$

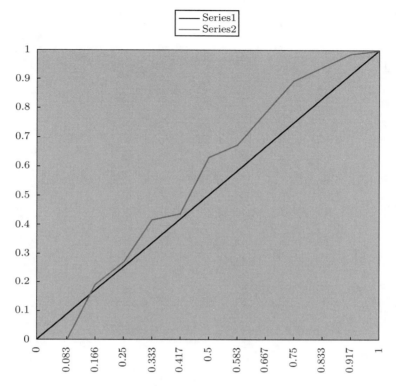

FIG. 2.2.2. *Scaled TTT plot for exponential random quantities (Series 1) and exponential transform (Series 2).*

corresponding to an exchangeable sample, $x_1, x_2, \ldots, x_n$. It is defined as

$$F_n(u) = \begin{cases} 0, & u < x_{(1)}, \\ \frac{i}{n}, & x_{(i)} \le u < x_{(i+1)}, \\ 1, & u \ge x_{(n)}. \end{cases} \qquad (2.2.2)$$

Since it is a step function, it is hardly a good estimator for a continuous life distribution. It is, in a sense, a nonparametric MLE for a univariate cumulative distribution corresponding to the exchangeable sample.

The scaled TTT plots correspond to a transform of the empirical cumulative distribution. There is likewise a corresponding transform for any cumulative probability distribution. For example, Figure 2.2.3 has graphs of transforms of Weibull distributions:

$$P(X \le x \,|\, \alpha, \beta) = F(x \,|\, \alpha, \beta) = 1 - e^{-(\frac{x}{\beta})^{\alpha}} \qquad (2.2.3)$$

for $x \ge 0$, α, $\beta > 0$. In this case β is a scale parameter while α governs the "shape" of the distribution and is therefore called a shape parameter.

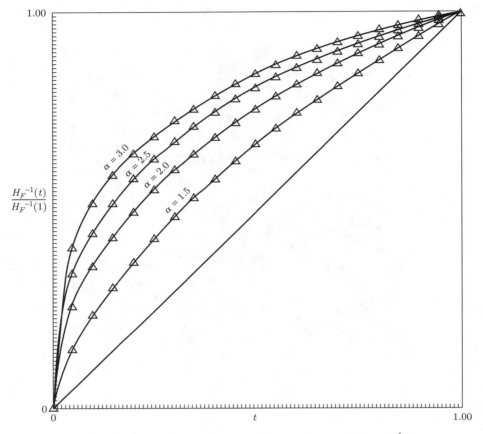

FIG. 2.2.3. *Scaled TTT transforms (Weibull distributions)* $(H_F^{-1}(t) = \int_0^{F^{-1}(t)} \overline{F}(u)\,du)$.

Comparison of Figures 2.2.1 and 2.2.3 suggests that perhaps a Weibull distribution could be used as a conditional probability model to analyze the data in Table 2.2.1. The shape parameter α could also be inferred but not the scale parameter β, since the plot is scale invariant.

The TTT transform. Suppose that we were to observe an ordered sample of observations

$$0 \equiv x_{(0)} \leq x_{(1)} \leq x_{(2)} \leq \cdots \leq x_{(r)}.$$

Suppose, furthermore, that in our judgment the distribution

$$G(x \mid \theta) = 1 - e^{-\frac{x}{\theta}}, \quad x \geq 0 \quad \text{and} \quad \theta > 0$$

is consistent with our prior knowledge.

As we showed in section 1, if we only observe the first r ordered values, then the likelihood is proportional to

$$\left[\prod_{i=1}^{r} e^{-\frac{x_{(i)}}{\theta}}\right] e^{-\frac{(n-r)t}{\theta}} = \frac{1}{\theta^r} e^{-\frac{T(t)}{\theta}},$$

where in this case $t = x_{(r)}$. The MLE for θ is

$$\hat{\theta} = \frac{1}{r} \left[\sum_{i=1}^{r} x_{(i)} + (n-r)x_{(r)}\right] \stackrel{DEF}{=} \frac{1}{r} T(x_{(r)}).$$

It is useful to observe that we can also write $T(x_{(r)})$ as

$$T(x_{(r)}) = nx_{(1)} + (n-1)(x_{(2)} - x_{(1)}) + \cdots + (n-r+1)(x_{(r)} - x_{(r-1)}), \quad (2.2.4)$$

the sum of all observed complete and incomplete lifetimes, or the TTT. This is also a very useful statistic for more general models.

Now suppose that the life distribution F $(F(0^-) = 0)$ is arbitrary. Another interpretation of $T(t)$ can be given in terms of the empirical distribution

$$F_n(u) = \begin{cases} 0, & u < x_{(1)}, \\ \frac{i}{n}, & x_{(i)} \le u < x_{(i+1)}, \\ 1, & u \ge x_{(n)}. \end{cases} \quad (2.2.5)$$

Define $F_n^{-1}(u) = \inf\{x \mid F_n(x) > u\}$ and note that

$$\int_0^{F_n^{-1}(\frac{r}{n})} \overline{F}(u)\, du = \sum_{i=1}^{r} \left[1 - \frac{(i-1)}{n}\right][x_{(i)} - x_{(i-1)}]$$

$$= \frac{1}{n} T(x_{(r)}). \quad (2.2.6)$$

The last equality follows from the expression for the TTT (2.2.4).

Assuming infinite exchangeability and conditional on F,

$$\lim_{\substack{n \to \infty \\ \frac{r}{n} \to t}} \int_0^{F_n^{-1}(\frac{r}{n})} \overline{F}_n(u)\, du = \int_0^{F^{-1}(t)} \overline{F}(u)\, du \quad (2.2.7)$$

uniformly in t $(0 \le t \le 1)$. We call

$$H_F^{-1}(t) = \int_0^{F^{-1}(t)} \overline{F}(u)\, du$$

the TTT transform. It is easy to verify that the mean, μ, of F is given by

$$H_F^{-1}(1) = \int_0^{F^{(-1)}(1)} \overline{F}(u)\, du = \mu.$$

As an example, let $G(x \mid \theta) = 1 - e^{-x/\theta}$. Then

$$H_G^{-1}(t) = \int_0^{G^{-1}(t)} e^{-\frac{x}{\theta}}\, dx = \int_0^{G^{-1}(t)} \theta\, dG(x) = \theta\, G[G^{-1}(t)] = \theta t$$

for $(0 \le t \le 1)$. Hence

$$\frac{H_G^{-1}(t)}{H_G^{-1}(1)} = t;$$

i.e., G is transformed into the 45° line on $[0, 1]$. We call $\frac{H_G^{-1}(t)}{H_G^{-1}(1)}$ the *scaled TTT transform*.

An important property of the transform is that

$$\left. \frac{d}{dt} H_F^{-1}(t) \right|_{t=F(x)} = \left. \frac{1-t}{f[F^{-1}(t)]} \right|_{t=F(x)} = \frac{1}{r(x)}, \qquad (2.2.8)$$

where

$$r(x) = \frac{f(x)}{\overline{F}(x)}$$

is the failure rate function at x. It follows from (2.2.8) that if $r(\bullet)$ is increasing, then $H_F^{-1}(t)$ is *concave* in t $(0 \le t \le 1)$.

Using (2.2.8) it is easy to see that H_F determines F if F is absolutely continuous. Since absolutely continuous F's are dense in the class of all failure distributions, we see that H_F determines F in general.

The scaled TTT statistic for exponentially distributed random quantities. Inference using scaled TTT plots is based on properties of the exponential distribution. Let $X_1, X_2, \ldots, X_n$ be n random quantities from $G(x \mid \theta = 1) = 1 - e^{-x}$ for $x \ge 0$. (We assume $\theta = 1$ for convenience only.) First note that G has the memoryless property; i.e.,

$$\overline{G}(t + x) = \overline{G}(t)\overline{G}(x)$$

for all t, $x \ge 0$. It follows that

$$P(X > t + x \mid X > t) = P(X > x). \qquad (2.2.9)$$

This is the so-called memoryless property of the exponential distribution. It is not true for the finite population exponential distribution.

THEOREM 2.2.1. *Let* $0 \equiv X_{(0)} \leq X_{(1)} \leq X_{(2)} \leq \cdots \leq X_{(r)} \leq \cdots \leq X_{(n)}$
be ordered random quantities from the distribution $G(x \mid \theta = 1) = 1 - e^{-x}$ *for*
$x \geq 0$. *Let*

$$D_i = X_{(i)} - X_{(i-1)}$$

for $i = 1, 2, \ldots, n$ *be the spacings between order statistics. Then*
(a) $P(D_i \leq t) = G_{(n-i+1)}(t) = 1 - e^{-(n-i+1)t}$ *for* $i = 1, 2, \ldots, n;$
(b) $D_1, \ldots, D_n$ *are mutually independent.*

Proof.
(a) First note that $P(D_1 > t) = e^{-nt}$, so that (a) holds for $i = 1$. By the memoryless property (2.2.9) of the exponential distribution, D_2 has the same marginal distribution as the first spacing D_1, but now from a sample of size $n - 1$ so that (a) holds for $i = 2$. By repeated applications of this argument, (a) holds for $i = 1, 2, \ldots, n$.

(b) Using the joint probability density for the order statistics, the joint probability density for the n spacings is given by

$$
\begin{aligned}
p(d_1, \ldots, d_n \mid \theta = 1) &= n! \prod_{i=1}^{n} e^{-(n-i+1)d_i} \\
&= \prod_{i=1}^{n} (n - i + 1) e^{-(n-i+1)d_i}. \qquad (2.2.10)
\end{aligned}
$$

Note that the ith factor $(n - i + 1) e^{-(n-i+1)d_i}$ represents the marginal density of D_i obtained in (a) above. Since the joint density of $D_1, \ldots, D_n$ factors into the product of the individual densities of the D_i, then the quantities $D_1, \ldots, D_n$ are mutually independent. □

It follows from Theorem 2.2.1 that since the normalized spacings

$$nD_1, (n-1)D_2, \ldots, (n-i+1)D_i, \ldots, D_n$$

are distributed with the exponential distribution $G(x \mid \theta = 1) = 1 - e^{-x}$, they are independently distributed with common exponential distribution. It follows that the TTT random quantity

$$T(X_{(r)}) = \sum_{i=1}^{r} (n - i + 1)(X_{(i)} - X_{(i-1)}) = \sum_{i=1}^{r} (n - i + 1) D_i$$

is distributed as the sum of r independent unit exponential distributions.

We next show that the distribution of $T(X_{(r)})$ is a gamma distribution with density

$$g_{a,b}(y) = \frac{b^a}{\Gamma(a)} y^{a-1} e^{-by}$$

for $y \geq 0$ with $a = r$ and $b = 1$.

THEOREM 2.2.2. *Let* $Y_1, \ldots, Y_n$ *be independently and identically distributed with exponential distribution* $G(x \mid \theta = 1) = 1 - e^{-x}$. *Then*

$$Z = Y_1 + \cdots + Y_r$$

has density

$$g_r(z) = \frac{z^{r-1}e^{-z}}{(r-1)!}. \tag{2.2.11}$$

Hence the density of $T(X_{(r)})$ *is a gamma density with shape parameter* r.

Proof. First note that (2.2.11) holds for $n = 1$. Assume (2.2.11) holds for $n = k$. Then for $z \geq 0$,

$$g_{k+1}(z) = \int_0^z g_k(x)e^{-(z-x)}dx,$$

where the integrand represents the joint probability density that

$$Y_1 + \cdots + Y_k = x,$$

$Y_{k+1} = z - x$, and the range $[0, z]$ of integration corresponds to the set of mutually exclusive and exhaustive outcomes for $Y_1 + \cdots + Y_k$. Hence

$$g_{k+1}(z) = \int_0^z \frac{x^{k-1}}{(k-1)!}e^{-(z-x)}dx = e^{-z}\int_0^z \frac{x^{k-1}}{(k-1)!}dx$$

$$= \frac{z^k}{k!}e^{-z}.$$

Thus (2.2.11) holds for $n = k+1$. By induction it follows that the theorem holds for $n = 1, 2, \ldots$. □

THEOREM 2.2.3. *Let* $0 \equiv X_{(0)} \leq X_{(1)} \leq X_{(2)} \leq \cdots \leq x_{(r)} \leq \cdots \leq X_{(n)}$ *be ordered random quantities from the distribution* $G(x \mid \theta = 1) = 1 - e^{-x}$ *for* $x \geq 0$ *and*

$$T(X_{(r)}) = \sum_{i=1}^r (n-i+1)(X_{(i)} - X_{(i-1)}).$$

Then

$$\frac{T(X_{(r)})}{T(X_{(n)})}, \qquad r = 1, 2, \ldots, n-1$$

are distributed as the order statistics from a uniform distribution on $[0,1]$.

The proof depends on the connection between the gamma and beta distributions. If Y and Z are independent gamma distributed random quantities with shape parameters r and $n - r$, respectively, then the quotient

$$\frac{Y}{Y + Z}$$

is a beta distributed random quantity with parameters r and $n - r$. This is also the density of the rth-order statistic from $n - 1$ independent, uniformly distributed random quantities on $[0, 1]$. For a generalized proof, see Chapter 3, section 3.2, Theorem 3.2.1.

Comments on graphical methods. Graphical methods for plotting lifetimes are commonly based on the cumulative empirical distribution (2.2.2). Items must first be judged exchangeable with respect to lifetime before making the graph. This cumulative empirical distribution is, in a sense, an *MLE* of a univariate distribution assuming the population size $N \uparrow \infty$. Since it is a step function, it is a poor estimator of distributions considered a priori continuous such as life distributions. We have suggested using transforms of the empirical distribution as a means of guessing appropriate conditional probability models, conditional on the scale parameter. The scale parameter must still be inferred based on bayesian methods.

It is important to realize that the preceding argument has been deductive in nature rather than inductive. That is, we began our mathematical argument with the cumulative distribution F specified and then deduced properties of a transform of this distribution. This is deductive analysis.

Based on data judged a priori exchangeable, we plotted the data as in Figure 2.2.1. From this data plot we then tried to make inferences about an unknown distribution F by comparing the data plot with transform plots of known distributions. Since samples are necessarily finite, F unknown, and our argument concerning properties of the transform of F are deductive in nature, we should not stop our statistical analysis at this point but go on to a full bayesian analysis using additional a priori engineering judgments.

Ideally we should derive a likelihood model using engineering knowledge and indifference arguments based on judgments such as exchangeability. In Chapter 4 we show how this can be done with respect to the strength of materials.

Exercises

2.2.1. Show that if the random quantity X has distribution $G(x \mid \theta = 1) = 1 - e^{-x}$ for $x \geq 0$ then $U \overset{DEF}{=} G(X \mid \theta = 1)$ has a uniform distribution on $[0, 1]$.

2.2.2. Generate $n = 20$ random exponentially distributed random quantities using a random number generator with a computer using the result in Exercise 2.2.1. Construct a TTT data plot using the 20 numbers generated and count the number of crossings with respect to the $45°$ line.

2.2.3. Construct a TTT data plot for the lifetimes of Kevlar 49/epoxy strands given in Table 2.2.3. The strands were put on life test at the 80% stress level, i.e., at 80% of mean rupture strength. Does the TTT plot suggest that an exponential model might be appropriate? If so, what calculations would you perform next?

TABLE 2.2.3

Lifetimes of Kevlar 49/epoxy strands loaded at 80% of mean rupture strength.

Rank	Time to failure	Rank	Time to failure	Rank	Time to failure	Rank	Time to failure
1	1.8	26	84.2	51	152.2	76	285.9
2	3.1	27	87.1	52	152.8	77	292.6
3	4.2	28	87.3	53	157.7	78	295.1
4	6	29	93.2	54	160	79	301.1
5	7.5	30	103.4	55	163.6	80	304.3
6	8.2	31	104.6	56	166.9	81	316.8
7	8.5	32	105.5	57	170.5	82	329.8
8	10.3	33	108.8	58	174.9	83	334.1
9	10.6	34	112.6	59	177.7	84	346.2
10	24.2	35	116.8	60	179.2	85	351.2
11	29.6	36	118	61	183.6	86	353.3
12	31.7	37	122.3	62	183.8	87	369.3
13	41.9	38	123.5	63	194.3	88	372.3
14	44.1	39	124.4	64	195.1	89	381.3
15	49.5	40	125.4	65	195.3	90	393.5
16	50.1	41	129.5	66	202.6	91	451.3
17	59.7	42	130.4	67	220.2	92	461.5
18	61.7	43	131.6	68	221.3	93	574.2
19	64.4	44	132.8	69	227.2	94	653.3
20	69.7	45	133.8	70	251	95	663
21	70	46	137	71	266.5	96	669.8
22	77.8	47	140.2	72	267.9	97	739.7
23	80.5	48	140.9	73	269.2	98	759.6
24	82.3	49	148.5	74	270.4	99	894.7
25	83.5	50	149.2	75	272.5	100	974.9

Time to failure in hours, $n = 100$.

2.2.4. Construct a TTT data plot of the breaking load strength data on Kevlar 49/epoxy strands in Table 2.2.4. What can you infer about the strength data?

2.2.5. Assume that random quantities are exponentially distributed. Use Theorem 2.2.3 to calculate

$$P\left(\frac{T(X_{(1)})}{T(X_{(n)})} > \frac{1}{n}\right),$$

i.e., the probability that the first point on the TTT plot lies above the 45° line. What is $\lim_{n\uparrow\infty} P(\frac{T(X_{(1)})}{T(X_{(n)})} > \frac{1}{n})$?

2.2.6. State and prove Theorems 2.2.1 and 2.2.2 for the case $G(x\,|\,\theta) = 1 - e^{-x/\theta}$; i.e., θ is not necessarily 1.

TABLE 2.2.4

Breaking load strength of individual Kevlar 49/epoxy strands.

Specimen number	Load lb.	Specimen number	Load lb.	Specimen number	Load lb.
1	22.0	19	21.8	37	21.1
2	22.0	20	22.9	38	22.2
3	22.0	21	21.8	39	21.4
4	21.6	22	23.1	40	22.5
5	21.6	23	22.9	41	21.4
6	22.0	24	22.0	42	21.6
7	22.0	25	22.2	43	22.0
8	20.9	26	22.5	44	21.4
9	22.5	27	21.4	45	21.4
10	22.7	28	22.2	46	22.7
11	22.9	29	22.7	47	21.6
12	22.0	30	22.7	48	21.6
13	22.5	31	22.5	49	22.7
14	22.5	32	22.2	50	21.8
15	21.6	33	22.5	51	22.0
16	21.8	34	22.0	52	22.2
17	22.0	35	20.5	53	22.0
18	22.0	36	22.5		

2.2.7*. Compute the distribution of $T(X_{(r)})$ when random quantities are from the finite population exponential distribution of Chapter 1 with N the population size, n the sample size, and r corresponding to the rth order statistic. (*Hint.* Compare with (1.1.4) of Chapter 1.)

2.3. Weibull analysis.

Perhaps the most widely used probability distribution in engineering reliability next to the exponential distribution is the Weibull distribution, named after a Swedish mechanical engineer by the name of Walodie Weibull [Weibull (1939)]. Weibull was primarily interested in the strength of materials. However, the distribution is just as often used in analyzing lifetime data. Let X be the lifetime of an item or device. The form of the Weibull *survival* distribution is

$$P(X > x \mid \lambda, \alpha) = \overline{F}(x \mid \lambda, \alpha) = e^{-(\lambda x)^{\alpha}} \quad \text{for } x \geq 0 \quad \text{and 1 for } x < 0 \quad (2.3.1)$$

with density

$$f(x \mid \lambda, \alpha) = \alpha \lambda^{\alpha} x^{\alpha-1} e^{-(\lambda x)^{\alpha}}.$$

Another function of interest in reliability is the failure rate function (sometimes called the "force of mortality"). If a cumulative probability distribution F has density f, then the *failure rate* function is

$$r(x) = \frac{f(x)}{\overline{F}(x)}. \qquad (2.3.2)$$

The failure rate for the Weibull distribution is

$$r(x \mid \lambda, \alpha) = (\alpha\lambda)(\lambda x)^{\alpha-1} \qquad (2.3.3)$$

for $x \geq 0$ and 0 otherwise. When $\alpha = 1$, the failure rate is constant, and the Weibull distribution reduces to the exponential distribution. For $\alpha > 1$ ($\alpha < 1$) the failure rate is increasing (decreasing). The mean of the Weibull distribution is

$$E(X \mid \lambda, \alpha) = \frac{1}{\lambda}\Gamma\left(1 + \frac{1}{\alpha}\right) \qquad (2.3.4)$$

while the variance is

$$Var(X \mid \lambda, \alpha) = \frac{1}{\lambda^2}\left(\Gamma\left(1 + \frac{2}{\alpha}\right) - \Gamma^2\left(1 + \frac{1}{\alpha}\right)\right). \qquad (2.3.5)$$

Sometimes three parameters are used where the third parameter ε is the smallest possible judged value of X.

Example 2.3.1. TOW engine failure data. High by-pass turbofan jet engine corrective removal data were given in Table 2.2.1. The removal rates of such engines are of interest in assessing direct maintenance costs and planning preventive removals. We wish to analyze the data in Table 2.3.1 on engine removal times including noncorrective removals (i.e., preventive removals). Weibull analysis is one way of doing this. The data is fairly "clean" except for the presence of survivals that were not failures. Note that most of the data are of this type. There were only 12 recorded "failures," i.e., service removals, out of the 29 observations.

The question we wish to answer is, How can we analyze data of this type?

Perhaps the most distinguishing property of the class of Weibull distributions is that they are preserved under the operation of taking the minimum. Thus if $X_1, X_2, \ldots, X_n$ are independent (conditional on λ and α) Weibull distributed random quantities with survival distribution (2.3.1), then

$$P(\min(X_1, X_2, \ldots, X_n) > x \mid \lambda, \alpha) = e^{-n(\lambda x)^\alpha},$$

where the new Weibull parameters are $n^{1/\alpha}\lambda$ and α. The connection with data analysis, however, is unclear!

Gnedenko (1943) showed that there are only three possible limiting distributions for n independent minima when the (arbitrary) survival distribution is normalized as

$$\overline{F}(\alpha_n x + \beta_n),$$

TABLE 2.3.1

A 1 in the third column indicates a service removal while a 0 indicates survival and no need for removal.

Engine shop visit removals		
Serial number	Flight hours	Number of removals
ABC123	389	1
ABC124	588	1
ABC125	665	0
ABC126	707	0
ABC127	715	0
ABC128	732	0
ABC129	768	0
ABC130	778	0
ABC131	796	0
ABC132	808	0
ABC133	840	1
ABC134	853	1
ABC135	868	1
ABC136	880	0
ABC137	931	1
ABC138	977	0
ABC139	997	1
ABC140	1020	0
ABC141	1060	1
ABC142	1074	0
ABC143	1091	0
ABC144	1122	0
ABC145	1153	0
ABC146	1200	1
ABC147	1258	1
ABC148	1337	1
ABC149	1411	1
ABC150	1467	0
ABC151	1829	0
TOTAL		29

where α_n and β_n are suitable norming constants depending on n and the particular F. One of these distributions (the only one with positive support) is the Weibull distribution. Again, it is not clear what, if any, connection exists with inferential data analysis.

The likelihood principle. If we judge $\overline{F}(x \mid \lambda, \alpha) = e^{-(\lambda x)^{\alpha}}$ as our probability model for observations conditional on (λ, α), then the likelihood $L(\lambda, \alpha)$ can in principle be calculated for any data set. By the *likelihood principle* [cf. D. Basu (1988)], the likelihood contains all the information the data can give us about the parameters. This has far-reaching consequences for the way in which data is analyzed, not only with respect to the Weibull distribution. But the Weibull distribution by its form makes the likelihood particularly easy to compute. However, before we calculate the likelihood for the type of data in Table 2.3.1, it will be convenient to calculate the likelihood for general failure rate functions

$$r(x) = \frac{f(x)}{\overline{F}(x)}.$$

The likelihood for general failure rate functions. Both the life distribution F and the density f can be represented in terms of the failure rate function $r(x) = \frac{f(x)}{\overline{F}(x)}$. To see this note that

$$\frac{d}{dx}\{-\ln \overline{F}(x)\} = \frac{f(x)}{\overline{F}(x)} = r(x)$$

so that

$$-\ln \overline{F}(x) = \int_0^x r(u)\, du$$

and

$$\overline{F}(x) = e^{-\int_0^x r(u)du}, \tag{2.3.6}$$

while

$$f(x) = r(x)e^{-\int_0^x r(u)\, du}. \tag{2.3.7}$$

The following result will be very useful in computing the likelihood in the Weibull case.

THEOREM 2.3.1. *Given the failure rate function $r(x)$, conditionally independent observations are made. Let $x_1, x_2, \ldots, x_k$ denote k observed failure ages (in Table 2.3.1, $k = 12$) and $\ell_1, \ell_2, \ldots, \ell_m$ the m survival ages (in Table 2.3.1, $m = 17$). The notation ℓ denotes loss, as lost from further observation. Let $n(u)$ denote the number of items or devices under observation at age u, $u \geq 0$, and let $r(u)$ denote the failure rate function of the item at age u. Then the likelihood corresponding to the failure rate function considered as a continuous parameter, having observed data, is given by*

$$L(r(u), u \geq 0),$$

$$\propto \begin{cases} \left[\prod_{s=1}^k r(x_s)\right] \exp[-\int_0^\infty n(u)r(u)du] & \text{when } k \geq 1, \\ \exp[-\int_0^\infty n(u)r(u)du] & \text{when } k = 0. \end{cases} \tag{2.3.8}$$

Proof. To justify this likelihood expression, we first note that the underlying random events are the ages at failure or withdrawal. Thus the likelihood of the observed outcome is specified by the likelihood of the failure ages and survival ages until withdrawal.

To calculate the likelihood, we use the fact that, given $r(\bullet)$,

$$f(x) = r(x) \exp \left[- \int_0^x r(u) du \right].$$

Specifically, if an item is observed from age 0 until it is withdrawn at age ℓ_t, without having failed during the interval $[0, \ell_t]$, a factor

$$\exp \left[- \int_0^{\ell_t} r(u) du \right]$$

is contributed to the likelihood. Thus, if no items fail during the test (i.e., $k = 0$), the likelihood of the observed outcome is proportional to the expression given for $k = 0$.

On the other hand, if an item is observed from age 0 until it fails at age x_s, a factor

$$r(x_s) \exp \left[- \int_0^{x_s} r(u) du \right]$$

is contributed to the likelihood. The exponential factor corresponds to the survival of the item during $[0, x_s]$, while $r(x_s)$ represents the rate of failure at age x_s. (Note that if we had retained the differential element "dx," the corresponding expression $r(x_s)dx_s$ would approximate an actual probability: the conditional probability of a failure during the interval $(x_s, x_s + dx_s)$ given survival to age x_s.)

The likelihood expression corresponding to the outcome $k \geq 1$ now is clear. The exponential factor corresponds to the survival intervals of both items that failed under observation and items that were withdrawn before failing:

$$\int_0^{\infty} n(u) r(u) du = \sum_{s=1}^{k} \int_0^{x_s} r(u) du + \sum_{t=1}^{m} \int_0^{\ell_t} r(u) du,$$

where the first sum is taken over items that failed while the second sum is taken over items that were withdrawn. The upper limit "∞" is for simplicity and introduces no technical difficulty, since $n(u) \equiv 0$ after observation ends. □

The likelihood we have calculated applies for any absolutely continuous life distribution. In the special case of an exponential life distribution model $f(x \mid \lambda) = \lambda e^{-\lambda x}$, the likelihood of the observed outcome takes the simpler form

$$L(\lambda) \propto \begin{cases} \lambda^k \exp \left[-\lambda \int_0^{\infty} n(u) \, du \right] & \text{when } k \geq 1, \\ \exp \left[-\lambda \int_0^{\infty} n(u) \, du \right] & \text{when } k = 0. \end{cases} \qquad (2.3.9)$$

Back to the Weibull example.   Although failures and losses will usually be recorded in calendar time, for the purpose of analysis we relate all ages from the "birth" date of the item or the time at which the item was put into operation.

Let $x_1, x_2, \ldots, x_k$ denote the unordered observed failure ages and $n(u)$ the number of devices surviving to age u. (Note that since observation must stop at some finite time, $n(u) = 0$ for u sufficiently large. Furthermore, $n(u)$ is a step function.) Using Theorem 2.3.1 we can compute the likelihood for the Weibull distribution, namely,

$$L(\alpha, \lambda) \propto \alpha^k \lambda^{k\alpha} \left[\prod_{i=1}^{k} x_i^{\alpha-1} \right] \exp \left\{ -\lambda^\alpha \left[\int_0^\infty \alpha n(u) u^{\alpha-1} du \right] \right\} \qquad (2.3.10)$$

for $x_i \geq 0$ and $\alpha, \lambda > 0$. Suppose there are m withdrawals and we pool observed failure and loss times and relabel them as

$$0 \equiv t_{(0)} \leq t_{(1)} \leq \cdots \leq t_{(k+m)} \leq t.$$

If observation is confined to the age interval $[0, t]$, then we have

$$\int_0^\infty n(u) u^{\alpha-1} du = \sum_{i=1}^{k+m} (n - i + 1) \int_{t_{(i-1)}}^{t_{(i)}} u^{\alpha-1} du + (n - k - m) \int_{t_{(k+m)}}^{t} u^{\alpha-1} du.$$
$$(2.3.11)$$

Two important deductions can be made from this.

(1) From the likelihood (2.3.10) we can see that the only sufficient statistic for all parameters in the case of Weibull distributions is the entire data set. In other words there is no lower-dimensional statistical summary of the data.

(2) No natural conjugate family of priors is available for all parameters. Consequently, the posterior distribution must be computed using numerical integration.

Using three-dimensional graphics, contour plots for $L(\lambda, \alpha)$ can be made corresponding to selected *probability regions*. Often .99, .95, .50 probability regions are used although there is no special reason for this selection.

Example 2.3.1 (*Continued*). TOW engine failure data.   A TTT plot of just the failure times of the engines given in Table 2.2.1 suggests that a Weibull distribution with $\alpha = 3$ could be used to analyze the complete set of failure times in Table 2.3.1. If we pursue a Weibull analysis using $\alpha = 3$, we can utilize the additional information given in Table 2.3.1. In this case the parameter of interest is λ.

The first calculation should be the MLE for λ conditional on α. Note that in this case $n = 29$,

$$\alpha \int_0^\infty n(u)u^{\alpha-1}du = \alpha \sum_{i=1}^n (n-i+1) \int_{t_{(i-1)}}^{t_{(i)}} u^{\alpha-1}du$$

$$= \sum_{i=1}^n t_i^\alpha \stackrel{DEF}{=} T^*,$$

where T^* is the *transformed TTT*. The likelihood for λ can be written as

$$L(\lambda) \propto \lambda^{k\alpha} \exp\{-\lambda^\alpha T^*\}.$$

Calculating the MLE for λ we find

$$\hat{\lambda} = \left[\frac{k}{T^*}\right]^{\frac{1}{\alpha}}.$$

Since the mean of the Weibull distribution is

$$\theta = \frac{\Gamma\left(1+\frac{1}{\alpha}\right)}{\lambda},$$

the MLE for the mean is

$$\hat{\theta} = \frac{\Gamma\left(1+\frac{1}{\alpha}\right)}{\hat{\lambda}}$$

when α is specified.

To determine accuracy of the MLE for λ we plot the ratio of the likelihood divided by the likelihood evaluated at $\hat{\lambda}$; i.e.,

$$\frac{L(\lambda)}{L(\hat{\lambda})}$$

for λ's in a suitably chosen interval about $\hat{\lambda}$. This is the modified posterior for λ corresponding to a flat prior.

Exercises

2.3.1. Let X have a Weibull distribution with parameters λ, α.
(a) Show that the distribution of X^α given λ and α are exponential.
(b) What is the mean and variance of X^α given λ, α?
2.3.2. Assume $\alpha = 3$. Use the data in Table 2.3.1 to calculate the MLE $\hat{\lambda}$. Graph

$$\frac{L(\lambda)}{L(\hat{\lambda})}$$

to assess the accuracy of $\hat{\lambda}$. What is the MLE for the mean of the Weibull distribution?
2.3.3. Show that (2.3.4) and (2.3.5) are the mean and variance, respectively, of the Weibull distribution.
2.3.4. Why is the entire data set the *only* sufficient statistic for both parameters λ and α in the case of the Weibull distribution?

2.3.5. Calculate the exact likelihood expression for $\{r(u), u \geq 0\}$. The proportional likelihood is given by (2.3.8).

2.3.6. Graph the scaled TTT

$$\frac{\int_0^{F^{-1}(t)} \overline{F}(u)\, du}{\mu} \quad \text{for } 0 \leq t \leq 1$$

when $\overline{F}(x \mid \lambda, \alpha = 2) = e^{-(\lambda x)^2}$, $x \geq 0$, $\lambda > 0$ and the mean of F is μ.

2.3.7. Graph the predictive density for lifetime using the data in Table 2.3.1 with $\alpha = 3$; i.e., graph

$$p(x \mid \text{data}, \alpha = 3) \propto \int_0^\infty \alpha \gamma x^{\alpha-1} e^{-\gamma x^\alpha} L(\gamma) d\gamma,$$

where $\gamma = \lambda^\alpha$. It is convenient to use the proportional likelihood for λ^α rather than the likelihood for λ since $\alpha = 3$.

2.4.   Notes and references.

Section 2.1. The material on the influence of failures on the posterior density was taken from the 1985 paper, "Inference for the Exponential Life Distribution" by Barlow and Proschan. The appendix to that paper contains interesting proofs of similar results for the exponential model and arbitrary prior distributions for θ.

Section 2.2. TTT plots were first described in a 1975 paper by Barlow and Campo, "Total Time on Test Processes and Applications to Failure Data Analysis." A 1977 paper by Bo Bergman derived the distribution for the number of crossings of the TTT plot in the case of the exponential distribution model. Further properties of the TTT transform were described in the 1979 paper by Barlow, "Geometry of the Total Time on Test Transform."

The TTT plotting technique fails when data is incomplete. In the complete data case, the argument for the technique rests on the assumption of exchangeability, a stochastic comparison with the exponential case and convergence for infinite populations. The argument is not inductive.

Section 2.3. The basis for the Weibull analysis presented here is the likelihood principle [cf. Basu (1988)].

The *likelihood principle*: If two experiments E_1 and E_2 resulting in corresponding data sets x_1 and x_2 generate equivalent likelihood functions for the parameter of interest, then the information provided by E_1 is equal to the information provided by E_2.

The likelihood functions are equivalent if they are equal up to a constant factor independent of the parameter of interest. Information is anything concerning the parameter of interest that changes belief, expressed as a probability, about the parameter.

The likelihood for general failure rate functions was described in the paper by Barlow and Proschan (1988), "Life Distribution Models and Incomplete Data."

Notation

$$\frac{\sigma}{\mu}$$

coefficient of variation; μ is the mean and σ is the standard deviation

$$p_{a,b}(\lambda) = \frac{b^a \lambda^{a-1}}{\Gamma(a)} e^{-b\lambda} \quad \lambda \geq 0$$

gamma density

$$p_{a,b}(\theta) = \frac{b^a \theta^{-(a+1)}}{\Gamma(a)} e^{-b\theta} \quad \theta \geq 0$$

inverted gamma density

$$p_{a,b}(x) = \frac{\Gamma(a+1)}{\Gamma(a)} \frac{b^a}{(x+b)^{a+1}} \quad x \geq 0$$

Pareto density

$$T(t) = \sum_{i=1}^{r}(n-i+1)(x_{(i)} - x_{(i-1)}) + (n-r)(t - x_{(r)})$$

TTT to time t

$$F_n(u) = \begin{cases} 0, & u < x_{(1)}, \\ \frac{i}{n}, & x_{(i)} \leq u < x_{(i+1)}, \\ 1, & u \geq x_{(n)}. \end{cases}$$

empirical cumulative distribution

$$F(x \mid \alpha, \beta) = 1 - e^{-\left(\frac{x}{\beta}\right)^{\alpha}}$$

Weibull cumulative distribution with shape parameter α and scale parameter β

$$H_F^{-1}(t) = \int_0^{F^{-1}(t)} \overline{F}(u)\, du$$

TTT transform

$$r(x) = \frac{f(x)}{\overline{F}(x)}$$

failure rate function

CHAPTER 3

Counting the Number of Failures

Often, the only failure information available for a particular item or device is the number of defectives in a given sample or the number that failed in a given time interval. In section 3.1 we derive conditional probability models for the number of defectives relative to both finite and infinite populations. In sections 3.2, 3.3, and 3.4 we consider conceptually infinite populations. In section 3.4 the number of failures may be observed in finite time intervals of length τ and relative to a finite time horizon T.

3.1. The hypergeometric and binomial distributions.

The hypergeometric and binomial are two of the most useful probability models relative to counting failures. The hypergeometric is determined by the exchangeability judgment relative to 0 or 1 random quantities (sometimes called indicator random quantities). However, this is not obvious and we prove it from first principles. The binomial is, in one case, an approximation based on a conceptually infinite exchangeable population of items. As noted elsewhere, exchangeability does not imply independence although iid does imply exchangeability. For exchangeability to imply independence, in the case of infinite populations, we need to specify the univariate marginal distribution. Except in the case of games of chance, this is not realistic. For example, suppose we randomly draw, with replacement, objects from an "urn" whose proportion is *known*. In this case, the drawing outcomes will be independent and the number drawn of one kind will have a binomial distribution. Were the proportion to be *unknown*, the outcomes of random drawings would not be independent. This is because, through drawing objects, we would gradually learn the value of the proportion. In this situation, drawing outcomes are dependent.

Indicator random quantities. Suppose items are labeled $\{1, 2, \ldots, N\}$. Let $x_i = 1$ if item i is defective and $x_i = 0$ otherwise. Suppose items are exchangeable with respect to whether or not they are defective. Of course, if we can visually inspect items before testing, we might not judge them exchangeable even though we might not know the test outcome for sure. Hence the judgment that items are exchangeable may not always be correct for us. The judgment of exchangeability is in fact a fairly strong judgment about items before they are tested. Of course, after testing, the items would no longer be considered exchangeable.

Let $\mathbf{x}_N = (x_1, x_2, \ldots, x_N)$ and the parameter of interest be

$$\sum_{i=1}^{N} x_i = S, \tag{3.1.1}$$

the number of defective items in the lot. Using this notation, $p(\mathbf{x}_N)$ is invariant under permutations of the coordinates; i.e.,

$$p(\mathbf{x}_N) = p(\mathbf{y}_N)$$

when $\mathbf{y}_N$ is a permutation of $\mathbf{x}_N$. Note that the principle of indifference holds; that is, all vectors $\mathbf{x}_N$ that sum to S are equally likely in our judgment. It follows that, conditional on S, the probability function is determined; i.e.,

$$p(\mathbf{x}_N \mid S, N) = \frac{1}{\binom{N}{S}}$$

when $\sum_{i=1}^{N} x_i = S$ and is 0 otherwise.

If S is unknown, the problem is to infer S based on a sample of size n. Under exchangeability and conditional on S, we obtain the *first derived* distribution

$$p(\mathbf{x}_n \mid S, N) = \frac{S(S-1)\cdots(S-s+1)(N-S)\cdots(N-S-n+s+1)}{N(N-1)\cdots(N-n+1)}, \tag{3.1.2}$$

where $s = \sum_{i=1}^{n} x_i$. Also

$$p(\mathbf{0}_n \mid S, N) = \frac{(N-S)\cdots(N-S-n+1)}{N(N-1)\cdots(N-n+1)},$$

$$p(\mathbf{1}_n \mid S, N) = \frac{(S)(S-1)\cdots(S-n+1)}{N(N-1)\cdots(N-n+1)}.$$

Note that $p(\mathbf{x}_n \mid S)$ remains exchangeable; i.e., any subsequence of an exchangeable sequence is exchangeable.

Since s and n are sufficient for S, we will only need the *second derived* distribution, the distribution for s conditional on S, namely, the hypergeometric distribution

$$p(s \mid S, n, N) = \frac{\binom{S}{s}\binom{N-S}{n-s}}{\binom{N}{n}} \tag{3.1.3}$$

for $s = 0, 1, \ldots, \min(n, S)$.

Letting $S, N \uparrow \infty$ in such a manner that $\frac{S}{N} \to \rho$, we have the binomial approximation to the hypergeometric; namely,

$$p(s \mid n, \rho) = \binom{n}{s} \rho^s (1-\rho)^{n-s} \tag{3.1.4}$$

FIG. 3.1.1. *Simple "influence diagram."*

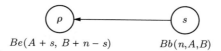

FIG. 3.1.2. *Reversing the arc in a simple influence diagram.*

for $s = 0, 1, \ldots, n$. The limiting proportion of defectives ρ is sometimes referred to as the "chance" that a specified item will be defective. A chance, however, is not the analyst's probability but an unknown parameter of the binomial model based on the concept of a limiting proportion (which is not really observable).

Note that in the limiting population case, any unexamined item would have the same a priori chance ρ of being defective irrespective of previous sample outcomes. That is, indicator random quantities would be independent, conditional on ρ.

Bayes' formula and the theorem of total probability. It is useful to represent random quantities by circles in a directed network (shown in Figure 3.1.1), where ρ is the parameter of a binomial distribution. The number of defectives s in a sample of size n has distribution $bi(n, \rho)$, and the arrow indicates that the distribution of s depends on ρ. In this situation we consider the sample size n known a priori. An unconditional distribution needs to be assigned to ρ. For convenience, suppose ρ is $Be(A, B)$.

Upon reversing the arrow, the distribution for ρ becomes a conditional distribution dependent on s. See Figure 3.1.2. The distribution of ρ has been updated by Bayes' formula to $Be(A + s, B + n - s)$; i.e.,

$$p_{A,B}(\rho \mid n, s) = \frac{\Gamma(A + B + n)}{\Gamma(A + s)\Gamma(B + n - s)} \rho^{A+s-1}(1 - \rho)^{B+n-s-1} \qquad (3.1.5)$$

for $0 \leq \rho \leq 1$.

On the other hand, the distribution of s is now unconditional. It is calculated using the "theorem of total probability" by merely unconditioning the original binomial distribution which was conditional on ρ. The unconditional distribution of s is now *beta-binomial* with parameters (n, A, B); i.e.,

$$
\begin{aligned}
p_{n,A,B}(s) &= \binom{n}{s} \frac{\Gamma(A + B)}{\Gamma(A)\Gamma(B)} \int_0^1 \rho^s(1 - \rho)^{n-s}\rho^{A-1}(1 - \rho)^{B-1} \, d\rho \\
&= \binom{n}{s} \frac{\Gamma(A + B)\Gamma(A + s)\Gamma(B + n - s)}{\Gamma(A + B + n)\Gamma(A)\Gamma(B)}
\end{aligned}
\qquad (3.1.6)
$$

$s = 0, 1, \ldots, n$.

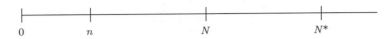

FIG. 3.1.3. *Relative sample and population sizes.*

The hypergeometric and the beta-binomial distributions. Suppose we have a unique lot of N exchangeable items; i.e., the population size is N. Let S be the number of items in the population that are defective. In order to learn about S, a sample of items of size n is examined and s, $0 \le s \le n$, are found to be defective. If we judge the items exchangeable with respect to being defective, then

$$p(s \mid S, n, N) = \frac{\binom{S}{s}\binom{N-S}{n-s}}{\binom{N}{n}}$$

for $s = 0, 1, \ldots, \min(n, S)$. Having observed s we wish to predict the number of defectives $S - s$ in the remainder consisting of $N - n$ items.

The natural conjugate prior for S in this case is the beta-binomial probability function, $Bb(N, A, B)$; see (3.1.6). The posterior probability function for $S - s$ given s is again beta-binomial but now with parameters $N - n, A + s, B + n - s$,

$$p_{N-n,A,B}(S - s \mid n, s)$$

$$= \int_0^1 \binom{N-n}{S-s} \rho^{S-s}(1-\rho)^{N-n-S-s} \qquad (3.1.7)$$

$$\cdot \frac{\Gamma(A+B+n)}{\Gamma(A+s)\Gamma(B+n-s)} \rho^{A+s-1}(1-\rho)^{B+n-s-1} \, d\rho.$$

To motivate the beta-binomial prior for S, suppose we fix N, S, s and let S^* be the number defective in $N^* > N$. See Figure 3.1.3. Now suppose $S^*, N^* \uparrow \infty$ in such a manner that $\frac{S^*}{N^*} \to \rho$. Then the number defective in N, namely, S, is binomially distributed with parameters (N, ρ). If we use the beta prior for ρ, namely, $Be(A, B)$, then the unconditional probability distribution for S given N

$$p_{N,A,B}(S)$$

will be the beta-binomial probability function given by (3.1.6) with N replacing n and S replacing s.

In the same way, $S - s$ given N, n, and ρ will be binomial with parameters $(N-n, \rho)$. Having observed s defectives in a sample of size n, we update the prior for ρ to $Be(A+s, B+n-s)$. Using this updated prior for ρ and the binomial for $S-s$ with parameters $(N-n, \rho)$, we obtain the posterior probability distribution for $S-s$ given n, s as in (3.1.7), namely, the $Bb(N-n, A+s, B+n-s)$ probability distribution.

Failure diagnosis. Off-shore oil pipelines on the seabed corrode with time and need to be tested for significant internal corrosion. This can be done using

FIG. 3.1.4. *Testing for corrosion.*

TABLE 3.1.1
Pipe corrosion test data.

State of pipe section	Test response		Sample size
	$t = 1$	$t = 0$	
Corroded	$x = 60$	90	$m = 150$
Not corroded	9	$y = 141$	$n = 150$

an instrument called a "pig" which is pushed through the pipe. The accuracy of the pig is to be determined and the probability calculated that a specified pipe section is significantly corroded when this is indicated by the pig.

Let

$$\delta = \begin{cases} 1 & \text{if there is significant pipe corrosion,} \\ 0 & \text{otherwise} \end{cases}$$

and

$$t = \begin{cases} 1 & \text{if the test indicates significant corrosion,} \\ 0 & \text{otherwise.} \end{cases}$$

We know that the test, although not perfect, depends on the state of the pipe section. Hence we want to reverse the arrow in Figure 3.1.4. But first, suppose we implement a planned experiment to determine the accuracy of the pig test. Suppose m pipe sections known to be corroded are tested and the test indicates that $x = 60$ of the m are corroded. Suppose also that n pipe sections known to be uncorroded are tested and the test indicates that $y = 141$ are uncorroded. Table 3.1.1 summarizes the experiment outcome.

Let π_i denote the proportion of times in a very large experiment that a corroded pipe section is correctly detected. Let

$$\pi_1 = P[t = 1 \mid \text{pipe section is corroded}].$$

Although we have used $P[\cdot]$ as if π_1 were a probability, this is an abuse of notation since π_1 is really a parameter, namely, a long run proportion (or chance).

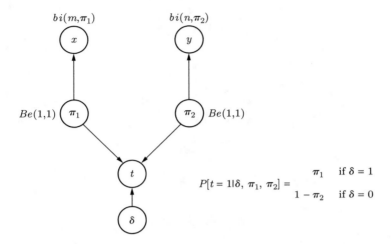

$$P[t = 1|\delta, \pi_1, \pi_2] = \begin{cases} \pi_1 & \text{if } \delta = 1 \\ 1 - \pi_2 & \text{if } \delta = 0 \end{cases}$$

FIG. 3.1.5. *Failure diagnosis influence diagram.*

Likewise, let

$$\pi_2 = P[t = 0 \mid \text{pipe section is not corroded}].$$

In engineering parlance, $1 - \pi_1$ is called the probability of a "false negative," while $1 - \pi_2$ is called the probability of a "false positive." In biometry, π_1 is called the "sensitivity" of the test, while π_2 is called the "specificity" of the test. We will extend Figure 3.1.4 to include this additional information; see Figure 3.1.5.

Suppose that initially $\pi_1 \sim Be(1,1)$ and also $\pi_2 \sim Be(1,1)$. Reversing the top arrows in Figure 3.1.5, $\pi_1 \mid x, m \sim Be(1 + x, 1 + m - x)$ and $\pi_2 \mid y, n \sim Be(1 + y, 1 + n - y)$. Eliminating π_1 and π_2 by taking expectations, $P(t = 1 \mid \delta = 1, x, m) = E(\pi_1 \mid x, m) = \frac{1+x}{2+m}$ while $P(t = 1 \mid \delta = 0, y, n) = E(1 - \pi_2 \mid y, n) = \frac{1+n-y}{2+n}$.

Figure 3.1.5 is now transformed into Figure 3.1.6. Suppose that a priori we judge

$$P(\delta = 1) = 0.10.$$

Reversing the arc from δ to t and using Bayes' formula we finally calculate

$$P[\delta = 1 \mid t, x, y] = \begin{cases} 0.10(\frac{x+1}{m+2})/P(t = 1 \mid x, y) & \text{if } t = 1, \\ 0.10(\frac{m-x+1}{m+2})/P(t = 0 \mid x, y) & \text{if } t = 0. \end{cases}$$

Using our tabulated values in Table 3.1.1, it follows that

$$P[\delta = 1 \mid t, x = 60, y = 141] = \begin{cases} 0.40 & \text{if } t = 1, \\ 0.066 & \text{if } t = 0. \end{cases}$$

The test has increased the probability of corrosion from 0.10 to 0.40 if $t = 1$.

Comments. In the 0-1 case, probability invariance with respect to permutations of vector components implies probability invariance with respect to sums

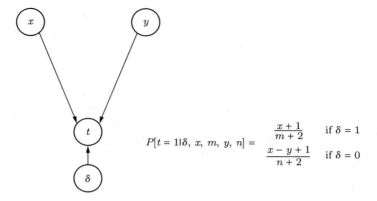

$$P[t = 1|\delta,\, x,\, m,\, y,\, n] = \begin{cases} \dfrac{x + 1}{m + 2} & \text{if } \delta = 1 \\[2mm] \dfrac{x - y + 1}{n + 2} & \text{if } \delta = 0 \end{cases}$$

FIG. 3.1.6. *"Instantiating" the data.*

of vector components and vice versa. In Chapter 1 this was not true. Although probability invariance with respect to sums does imply probability invariance with respect to permutations of vector components, the converse is false in general.

Exercises

3.1.1. Failure diagnosis. Pipes on the sea floor are tested for significant internal corrosion using a magnetic flux pig. The accuracy of the pig is to be determined and the probability calculated that a specified pipe section is significantly corroded when this is indicated by the pig.

Let

$$\delta = \begin{cases} 1 & \text{if there is significant corrosion,} \\ 0 & \text{otherwise,} \end{cases}$$

and

$$\delta = \begin{cases} 1 & \text{if the test indicates significant corrosion,} \\ 0 & \text{otherwise.} \end{cases}$$

Let π_1 denote the proportion of times in a very large experiment that correctly detects a corroded pipe section. Let

$$\pi_1 = P[t = 1 \mid \text{pipe section is corroded}].$$

Likewise, let

$$\pi_2 = P[t = 0 \mid \text{pipe section is not corroded}].$$

Suppose $n = 10$ pipe sections on the sea floor are tested and the tests indicate that all n are corroded. An "autopsy" is performed on each of the pipes and it is found that $x = 2$ are *not* significantly corroded.

(a) If a priori $\pi_1 \sim Be(5,1)$ and $\pi_2 \sim Be(1,1)$, what are the posterior distributions of π_1 and π_2?

(b) Let δ denote the indicator for a different pipe section not yet tested. If a priori $P(\delta = 1) = 0.10$, what is $P[\delta = 1 \mid t = 1, x, n]$?

3.1.2. Inspection sampling. An inspector records $x = 2$ defectives from a sample but forgets to record the sample size n. The sample is from a production process considered to be in statistical control; i.e., observations are considered to be exchangeable and as a consequence each item has the same a priori chance π of being defective. The inspector remembers that n is either 10 or 12 with equal probability. Suppose π is judged to have a $Be(1, 1)$ probability distribution.

(a) Draw an influence diagram containing nodes π, x, and n by adding the appropriate arrows together with the appropriate conditional probabilities. Using arc reversals, calculate $p(n \mid x, \pi)$.

(b) Using the influence diagram and arc reversals find a computing expression for $p(\pi \mid x)$.

3.1.3. The problem of two designs. There are two possible designs for constructing columns to be used in buildings. Each design involves the same construction cost. Let $\pi_1(\pi_2)$ be the "chance" of failure of a column due to a large earthquake if design one (two) is used. Suppose some tests have been performed using a shaker table (an earthquake simulator) and $p(\pi_1)[p(\pi_2)]$ is our probability function for $\pi_1(\pi_2)$. Suppose in addition that $E(\pi_1) = E(\pi_2)$ but $Var(\pi_1) << Var(\pi_2)$.

In each case below, let θ_i be the number of columns that fail in an earthquake if you use design $i(i = 1, 2)$. What is the probability function for $\theta_i(i = 1, 2)$ if columns fail independently given the earthquake occurs?

(a) Suppose you are considering building a structure requiring only one column. If an earthquake occurs and the column fails, you lose \$1 million and nothing otherwise. Which design would you choose?

(b) Suppose you are considering building another structure requiring two columns. If either or both columns fail in an earthquake, you lose \$1 million and nothing otherwise. Which design would you choose? (Assume columns fail independently given the occurrence of an earthquake.)

(c) Suppose you are considering constructing yet another building requiring three columns. Suppose π_1 is $Be(4, 6)$ and π_2 is $Be(2, 3)$ so that $E(\pi_1) = E(\pi_2)$ but $Var(\pi_1) < Var(\pi_2)$. If any columns fail in an earthquake, you lose \$1 million and nothing otherwise. Which design would you choose?

3.1.4*. Conditional dependence and unconditional independence in the finite population binary case. Let $S = \sum_{i=1}^{N} x_i$ and $s = \sum_{i=1}^{n} x_i$ where x_i is either 0 or 1 and

$$p(\mathbf{x}_n \mid S, N) = \frac{S(S-1)\cdots(S-s+1)(N-S)\cdots(N-S-n+s+1)}{N(N-1)\cdots(N-n+1)}.$$
(3.1.2)

Show that

$$\sum_{S=s}^{N-(n-s)} p(\mathbf{x}_n \mid S, N)p(S) = \prod_{i=1}^{n} \rho^{x_i}(1-\rho)^{1-x_i}$$

if and only if $p(S)$ is $bi(N, \rho)$. In this case *conditionally dependent* random quantities $X_1, X_2, \ldots, X_n$ with conditional density given by (3.1.2) are *unconditionally independent* when the prior distribution for S is binomial with parameters N (the population size) and ρ specified.

3.1.5*. Conditional dependence and unconditional independence in the finite population exponential case. Let $S = \sum_{i=1}^{N} x_i$ and show that

$$\prod_{i=1}^{n} \lambda e^{-\lambda x_i} = \int_{S=\sum_{i=1}^{n} x_i}^{\infty} \frac{(N-1)\cdots(N-n)}{S^n} \left[1 - \frac{\sum_{i=1}^{n} x_i}{S}\right]^{N-n-1} p(S)\, dS$$

if and only if $p(S \mid N, \lambda) = \frac{\lambda^N S^{N-1} e^{-\lambda S}}{\Gamma(N)}$. In this case *conditionally dependent* random quantities $X_1, X_2, \ldots, X_n$ with conditional density given by (1.1.1) in Chapter 1 are *unconditionally independent* when the prior distribution for $S = N\theta$ is a gamma density with shape parameter N (the population size) and λ specified.

3.2. Categorical data analysis.

In this section we consider exchangeable items which may be subjected to different tests and/or which may exhibit more than one failure mode. The population size will be considered conceptually infinite. For example, we could present items one after another to two certifiers, certifier 1 and certifier 2, to determine whether the two certifiers give substantially the same results. Each certifier classifies an item as top grade or not top grade. Fifty tests are recorded. The population of such items is considered unbounded, i.e., $N \uparrow \infty$. The results of the tests are shown in Table 3.2.1.

Corresponding to Table 3.2.1, consider the "chance" matrix in Table 3.2.2. What we call "chances" are not "probabilities" in our sense of probability. They are limiting proportions or so-called parameters. Although convenient concepts,

TABLE 3.2.1
Comparison of two certifiers.

Certifier 2	Certifier 1		
	Top grade	Not top grade	
Top grade	$n_{11} = 21$	$n_{12} = 8$	$n_{1\bullet} = 29$
Not top grade	$n_{21} = 3$	$n_{22} = 18$	$n_{2\bullet} = 21$
	$n_{\bullet 1} = 24$	$n_{\bullet 2} = 26$	$n_{\bullet\bullet} = 50$

TABLE 3.2.2
Chance matrix of two certifiers.

Certifier 2	Certifier 1	
	Top grade	Not top grade
Top grade	π_{11}	π_{12}
Not top grade	π_{21}	π_{22}

they can never be operationally observed since they involve infinite populations. In Table 3.2.2, π_{11} is the chance that a future item, exchangeable with the recorded items, will be given the designation top grade by both certifier 1 and 2. We are interested in estimating $\pi_{11} + \pi_{22}$, the chance that *both* certifiers will agree that a given item is either top grade or not top grade. To do this we need to first assign a joint prior probability function to $(\pi_{11}, \pi_{12}, \pi_{21}, \pi_{22})$. Note, however, that since $\pi_{11} + \pi_{12} + \pi_{21} + \pi_{22} = 1$, only three chances are free variables.

The Dirichlet probability distribution. Since, in general, categorical tables can be based on more than two categories, it will be convenient to use a different notation for the "chance" parameters.

Let $k + 1(k \geq 1)$ be the number of categories and let $Y_i(i = 1, 2, \ldots, k+1)$ be the chance that an item will fall in the category labelled i. In the example above we had two classifications and also two categories. We let π_{ij} be the chance that an item would fall in the category labelled (i, j). Instead of $(\pi_{11}, \pi_{12}, \pi_{21}, \pi_{22})$, we now use the notation (Y_1, Y_2, Y_3, Y_4) because this notation will be convenient when considering more than two classifications.

DEFINITION. *Let $\alpha_i > 0$ $(i = 1, 2, \ldots, k + 1)$. The random quantities $(Y_1, Y_2, \ldots, Y_k)$ are said to have a* Dirichlet *distribution with parameters $(\alpha_1, \alpha_2, \ldots, \alpha_{k+1})$ denoted by*

$$(Y_1, Y_2, \ldots, Y_k) \approx D_k(\alpha_1, \alpha_2, \ldots, \alpha_{k+1})$$

if the joint probability distribution of $(Y_1, Y_2, \ldots, Y_k)$ has joint probability density function

$$p(y_1, y_2, \ldots, y_k) = \frac{\Gamma(\alpha_1 + \alpha_2 + \cdots + \alpha_{k+1})}{\Pi_{i=1}^{k+1}\Gamma(\alpha_i)} y_1^{\alpha_1-1} y_2^{\alpha_2-1} \cdots y_k^{\alpha_k-1}$$

$$\cdot (1 - y_1 - y_2 - \cdots - y_k)^{\alpha_{k+1}-1} \qquad (3.2.1)$$

over the k-dimensional simplex defined by the inequalities $y_i > 0$, $i = 1, 2, \ldots, k$, $\sum_{i=1}^{k} y_i < 1$. Note that since $Y_{k+1} = 1 - Y_1 - Y_1 - \cdots - Y_k, Y_{k+1}$ is determined by $(Y_1, Y_2, \ldots, Y_k)$.

More generally, in this definition we can take $\alpha_i \geq 0$ for each i, and $\sum_{i=1}^{k+1} \alpha_i > 0$. If $\alpha_i = 0$ for some i then the corresponding $Y_i \equiv 0$.

For $k = 1$, the Dirichlet distribution $D_1(\alpha_1, \alpha_2)$ for Y_1 is the familiar beta distribution $Be(\alpha_1, \alpha_2)$ with parameters α_1 and α_2.

The multinomial probability distribution. Since, in general, categorical tables can be based on more than two categories, it will also be convenient to use a different notation for the number of observations in a given category.

Let $k + 1(k \geq 1)$ be the number of categories, and let $X_i(i = 1, 2, \ldots, k + 1)$ be the number of items that fall in the category labelled i. In the example above we had two classifications and also two categories. We let n_{ij} be the number of items that fell in the category labeled (i, j). Instead of $(n_{11}, n_{12}, n_{21}, n_{22})$, we now use the notation (X_1, X_2, X_3, X_4) because this notation will be convenient when considering more than two classifications.

DEFINITION. *The random quantities* $(X_1, X_2, \ldots, X_{k+1})$ *are said to have a multinomial probability distribution function with chance "probabilities"* $(y_1, y_2, \ldots, y_{k+1}) = \mathbf{y}_{k+1}$ *denoted by*

$$(X_1, X_2, \ldots, X_{k+1}) \mid n, \mathbf{y}_{k+1} \approx M(n; \mathbf{y}_{k+1})$$

if the joint probability function is

$$P(X_1 = x_1, X_2 = x_2, \ldots, X_{k+1} = x_{k+1} \mid Y_1 = y_1, Y_2 = y_2, \ldots, Y_{k+1} = y_{k+1})$$

$$= \frac{n!}{x_1! x_2! \ldots x_{k+1}!} y_1^{x_1} y_2^{x_2} \cdots y_{k+1}^{x_{k+1}}$$

(3.2.2)

for $x_i \geq 0, i = 1, 2, \ldots, k+1$ *where* $\sum_{i=1}^{k+1} x_i = n$. *Note that the multinomial coefficient* $\frac{n!}{x_1! x_2! \ldots x_{k+1}!}$ *is just the total number of different ways in which n items can be arranged when there are x_i items of type i* $(i = 1, 2, \ldots, k+1)$.

The posterior Dirichlet distribution.

From the definitions of the multinomial and Dirichlet distributions it follows that

$$p(y_1, y_2, \ldots, y_k \mid x_1, x_2, \ldots, x_{k+1}) \propto \left[\prod_{i=1}^{k} y_i^{\alpha_i + x_i - 1} \right] (1 - y_1 - y_2 - \cdots - y_k)^{\alpha_{k+1} + x_{k+1} - 1}.$$

(3.2.3)

It can be shown that the sum $Y_1 + Y_2 + \cdots + Y_m$ given $x_1, x_2, \ldots, x_{k+1}$ has a

$$Be \left(\sum_{i=1}^{m} (\alpha_i + x_i), \sum_{i=m+1}^{k+1} (\alpha_i + x_i) \right)$$

(3.2.4)

probability density. Using this result we can finally answer the question posed at the beginning of this section. Namely, what is the posterior probability distribution of $\pi_{11} + \pi_{22}$, the chance that both certifiers will give the same classification to a future item exchangeable with the ones observed?

If we use an initial Dirichlet prior $D_3(\alpha_{11}, \alpha_{12}, \alpha_{21}, \alpha_{22})$, the answer is

$$\pi_{11} + \pi_{22} \mid n_{11}, n_{12}, n_{21}, n_{22}$$

$$\approx Be(\alpha_{11} + \alpha_{22} + n_{11} + n_{22}, \alpha_{12} + \alpha_{21} + n_{12} + n_{21})$$

and

$$E[\pi_{11} + \pi_{22} \mid n_{11}, n_{12}, n_{21}, n_{22}] = \frac{\alpha_{11} + \alpha_{22} + n_{11} + n_{22}}{\alpha_{11} + \alpha_{22} + \alpha_{12} + \alpha_{21} + n_{11} + n_{22} + n_{12} + n_{21}}.$$

Recall that a $Be(A, B)$ probability density has mean $= \frac{A}{A+B}$ and variance $= \frac{AB}{(A+B)^2(A+B+1)}$. If we let $\alpha_{11} = \alpha_{12} = \alpha_{21} = \alpha_{22} = 1$ corresponding to a

TABLE 3.2.3
Missing failure mode data.

	Failure mode reported	Failure mode not reported
Ruptured	n_{11}	
		n_2
Leaking	n_{21}	
	$n_{\bullet 1}$	$n_{\bullet 2} = n_2$

TABLE 3.2.4
Chance table for data Table 3.2.3.

	Failure mode reported	Failure mode not reported
Ruptured	π_{11}	π_{12}
Leaking	π_{21}	π_{22}

"reference prior," then using the data in Table 3.2.1

$$E[\pi_{11} + \pi_{22} \mid n_{11}, n_{12}, n_{21}, n_{22}] = 0.79$$

and the standard deviation is

$$STD(\pi_{11} + \pi_{22} \mid n_{11}, n_{12}, n_{21}, n_{22}) = 0.016$$

so that there would not be much point in graphing the probability density.

Missing data. We will illustrate how partial information can be utilized to analyze failure data containing information on the mode of failure. Suppose that failure data are available for underground pipes that have been in place for some 30 years. In our example, pipe failure can be classified as either ruptured or leaking. Suppose in addition that some pipes known to have failed were not classified by failure mode. How can we make use of all the data to predict the proportion of ruptured pipes in the general population of such pipes? Table 3.2.3 illustrates the type of data under consideration.

The corresponding "chance" table is Table 3.2.4.

In Table 3.2.4, π_{12} is the proportion of ruptured pipes in an exchangeable infinite population of pipes whose failure mode would not be reported. We are interested in predicting $\pi_{1\bullet} = \pi_{11} + \pi_{12}$, which is similar to the previous example except for the missing data.

Again we will use a Dirichlet prior for $(\pi_{11}, \pi_{12}, \pi_{21})$ with parameters $(\alpha_{11}, \alpha_{12}, \alpha_{21}, \alpha_{22})$ where $\pi_{22} = 1 - \pi_{11} - \pi_{12} - \pi_{21}$. The trick is to make an "appropriate" change of variable with respect to the chances. Let

$$\theta = \pi_{11} + \pi_{21},$$

$$\theta_{11} = \frac{\pi_{11}}{\theta},$$

$$\theta_{12} = \frac{\pi_{12}}{1 - \theta},$$

so that the random quantity of interest with respect to this change of variables is

$$\pi_{1\bullet} = \pi_{11} + \pi_{12} = \theta\theta_{11} + (1 - \theta)\theta_{12}.$$

Next, it can be shown that

$$\theta \approx Be(\alpha_{\bullet 1}, \alpha_{\bullet 2}),$$

$$\theta_{11} \approx Be(\alpha_{11}, \alpha_{21}),$$

$$\theta_{12} \approx Be(\alpha_{12}, \alpha_{22}),$$

and that θ, θ_{11}, and θ_{12} are independent. The likelihood in terms of the new variables is

$$L = \theta^{n_{\bullet 1}}(1 - \theta)^{n_2}\theta_{11}^{n_{11}}(1 - \theta_{11})^{n_{21}}.$$

By matching the Dirichlet prior to this likelihood we obtain the posterior distribution of $(\theta, \theta_{11}, \theta_{12})$. It can be shown that θ, θ_{11}, and θ_{12} are also independent a posteriori and

$$\theta \mid (n_{11}, n_{21}, n_2) \approx Be(\alpha_{\bullet 1} + n_{\bullet 1}, \alpha_{\bullet 2} + n_{\bullet 2}),$$

$$\theta_{11} \mid (n_{11}, n_{21}, n_2) \approx Be(\alpha_{11} + n_{11}, \alpha_{21} + n_{21}),$$

$$\theta_{12} \mid (n_{11}, n_{21}, n_2) \approx Be(\alpha_{12}, \alpha_{22}).$$

It follows that

$$E(\pi_{1\bullet} \mid n_{11}, n_{21}, n_2) = E(\theta\theta_{11} + (1 - \theta)\theta_{12} \mid n_{11}, n_{21}, n_2)$$

$$= \frac{1}{\alpha + n}\left(\alpha_{1\bullet} + n_{11} + \frac{\alpha_{12}}{\alpha_{\bullet 2}}n_2\right),$$

where $n = n_{11} + n_{21} + n_2$. However, it would be much more difficult to calculate the posterior density of $\pi_{1\bullet} = \pi_{11} + \pi_{12}$.

The relationship between Dirichlet distributed random quantities and independent gamma distributed random quantities. The Dirichlet distribution can also be characterized in terms of mutually independent gamma random quantities. Recall that a random quantity Z has a gamma density $g(z)$

if

$$g(z) = \frac{b^\alpha z^{\alpha-1} e^{-bz}}{\Gamma(\alpha)}$$

for $z \geq 0, \alpha, b > 0$.

THEOREM 3.2.1. *Let $Z_1, Z_2, \ldots, Z_{k+1}$ be mutually independent gamma distributed random quantities with the common parameter $b > 0$ but with possibly different shape parameters $\alpha_i > 0$, $i = 1, 2, \ldots, k + 1$. Let $Z = \sum_{i=1}^{k+1} Z_i$ and $Y_i = \frac{Z_i}{Z}$. Then*

$$(Y_1, Y_2, \ldots, Y_k) \approx D_k(\alpha_1, \alpha_2, \ldots, \alpha_{k+1}).$$

Also $(Y_1, Y_2, \ldots, Y_k)$ are independent of Z.

Proof. The joint probability density function of the Z_i's is

$$p_{\alpha,b}(z_1, z_2, \ldots, z_{k+1}) \propto e^{-b\sum_{i=1}^{k+1} z_i} \prod_{i=1}^{k+1} z_i^{\alpha_i-1}$$

$z_i > 0$, $i = 1, 2, \ldots, k + 1$.

Consider the transformation

$$z = \sum_{i=1}^{k+1} z_i, \qquad y_i = \frac{z_i}{z},$$

$i = 1, 2, \ldots, k + 1$, the reverse transformation being $z_i = z y_i$, $i = 1, 2, \ldots, k + 1$ and $z_{k+1} = z(1 - \sum_{i=1}^{k} y_i)$.

The Jacobian

$$\left| \frac{\partial(z_1, z_2, \ldots, z_{k+1})}{\partial(z, y_1, \ldots, y_k)} \right| = z^k.$$

It follows then that the joint probability density function of $(Z, Y_1, \ldots, Y_k)$ is

$$g(z, y_1, \ldots, y_k) \propto e^{-bz} z^{\alpha-1} \prod_{i=1}^{k} y_i^{\alpha_i-1} \left(1 - \sum_{i=1}^{k} y_i\right)^{\alpha_{k+1}-1}. \qquad \square$$

When $\alpha_1 = \alpha_2 = \cdots = \alpha_{k+1} = 1$, $Z_1, Z_2, \ldots, Z_{k+1}$ are independent, exponentially distributed and $(Y_1, \ldots, Y_k)$ are independent uniform [0,1] random quantities. This proves Theorem 2.2.3 in Chapter 2.

Exercises

3.2.1. A cheap method versus a master method. Exchangeable items can be tested nondestructively using either a cheap method or a more expensive master method. The following table gives the data from a test wherein each of 40 items were tested with both methods.

	Master method	
Cheap method	Go	No go
Go	23	6
No go	5	7

Using the same Dirichlet prior employed for the example with two certifiers, calculate the expected proportion of times both methods would agree for a conceptually infinite population exchangeable with the tested items.

3.2.2. Some failure modes not reported. Calculate the expected proportion of ruptured pipes in a conceptually infinite population exchangeable with the items in the following table.

	Failure mode reported	Failure mode not reported
Ruptured	1	
		6
Leaking	3	
	4	6

Using the same Dirichlet prior as used for the missing data example, calculate the standard deviation of the posterior distribution for $\pi_{1\bullet} = \pi_{11} + \pi_{12}$.

3.2.3*. Sum of Dirichlet random quantities. If $(Y_1, Y_2, \ldots, Y_k) \approx D_k(\alpha_1, \alpha_2, \ldots, \alpha_{k+1})$, show that $\sum_{i=1}^{m} Y_i$ has a

$$Be\left(\sum_{i=1}^{m} \alpha_i, \sum_{i=m+1}^{k+1} \alpha_i\right)$$

distribution. (*Hint.* Similar to Exercise 1.1.4.)

3.3. Applications of the Poisson process.

In this section, the population size and the time horizon are both conceptually infinite. We start with the Poisson process model and show how it can be applied to practical problems.

A random process (or stochastic process) is a collection of random quantities, say, $\{X(t_1), X(t_2), \ldots, X(t_n)\}$, at time points (or epochs) $t_1 < t_2 < \cdots < t_n$. In our modeling leading to the Poisson process in section 3.4, time intervals over which failures are recorded will originally have length τ (e.g., Figure 3.3.1), where $T = K\tau$ is the time horizon. Consider the arbitrary time point t fixed.

The following two examples are illustrative of the kinds of failure data we have in mind. We will analyze these examples using the Poisson process approximation (corresponding to $T \uparrow \infty$ and $\tau \downarrow 0$). For the Poisson process approximation, let

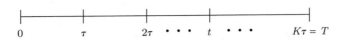

0   τ 2τ $\cdots$ t $\cdots$ $K\tau = T$

FIG. 3.3.1. *Time intervals for data records.*

TABLE 3.3.1
(M denotes million seeks.)

Hard disk failures		
	Number of seeks	Hard disk errors
Company A	250M	2500
Company B	1M	5

$N(t)$ denote the number of failures in $[0, t]$ while

$$P[N(t) = k \mid \lambda, t] = \frac{(\lambda t)^k e^{-\lambda t}}{k!} \qquad (3.3.1)$$

for $k = 0, 1, \ldots$ is the probability that k failures occur in $[0, t]$ conditional on the failure rate λ. Time can be measured in a number of ways. For example, automotive failures might be recorded relative to the number of miles driven. In the first example below, time is measured by the number of seeks of a computer hard disk. In the second example, time is measured by the number of laser shots.

Example 3.3.1. Hard disk failures. Two hard disk manufacturers were competing for a very large government contract. One condition of the competition was that the winner should have a lower failure rate with respect to hard disk failures. In Table 3.3.1 we have the following data based on field experience for many hard disks. Company A claimed that since they had many more disks in use, the apparent superiority of Company B was void. Would you agree? How would you analyze this data? The number of seeks is a measure of time passage in this case. We have recorded the total number of seeks over all hard disks of a particular type.

Example 3.3.2. Failure data for laser system components. A prototype laser system designed to generate and deliver high-power optical pulses to a target chamber to create fusion had the failure data for selected components as shown in Table 3.3.2.

The fact that some components had *no* recorded failures should be good news. However, the maximum likelihood estimate (MLE) for the failure rate in these cases would be 0, which is not reasonable, since we actually would expect failures at some future time. Time in this case is measured in the number of shots used for nuclear physics experiments.

TABLE 3.3.2

# Failures = k	# in system = n	Description
3	32	main bank charging supply
6	288	fuse
0	288	capacitor
0	512	flashlamp

Failures in 700 shots

TABLE 3.3.3

# Failures	TTT	MLE/1000 shots
3	22,400	0.13
6	201,600	0.03
0	201,600	0
0	358,400	0

Failures in 700 shots

In both examples, the time horizons T in terms of the number of seeks in the first example or the number of shots in the second example are very large. The total time on test (TTT) for all hard disks in the field in the first example is 250 million seeks for Company A and 1 million seeks for Company B. In the second example, we multiply the number of shots, 700, by the number of such devices in the system. Table 3.3.3 gives the MLEs per 1000 shots for the data in Table 3.3.2.

The Poisson probability model for a single component position. Using the derivation that follows in section 3.4, we can argue that the Poisson process provides a convenient probability model approximation for analyzing this failure data. Assume that the item in a given component position in each case is replaced by an exchangeable item at failure so that the population of such items is conceptually infinite. Let $N(t)$ be the number of failures in a *single* component position in time t. Let λ be the failure rate. Then

$$P[N(t) = k \mid \lambda, t] = \frac{(\lambda t)^k e^{-\lambda t}}{k!}, \tag{3.3.2}$$

$k = 0, 1, \ldots$ is the Poisson probability function approximation conditional on the failure rate λ. This model is a useful probability approximation for analyzing both data sets. The MLE for λ in (3.3.2) is

$$\hat{\lambda} = \frac{k}{t},$$

where k is the number of observed failures and t corresponds to the system operating time.

The posterior density for λ (a single component position). Consider data for items in a single component position which has k recorded failures in $[0, t]$. This can occur if the item is replaced at failure by exchangeable items. A useful calculation would be the posterior density for λ conditional on the data. Using Bayes' formula we have

$$p(\lambda \mid k, t) \propto \frac{(\lambda t)^k e^{-\lambda t}}{k!} p(\lambda).$$

In the absence of further information we could take

$$p(\lambda) = \text{ constant.}$$

Remembering the form of the gamma density

$$g_{a,b}(x) = \frac{b^a x^{a-1} e^{-bx}}{\Gamma(a)}$$

for $x \geq 0$, $a, b > 0$, we calculate

$$p(\lambda \mid k, t) = \frac{t^{k+1} \lambda^k e^{-\lambda t}}{k!} \qquad (3.3.3)$$

for $\lambda \geq 0$, where we have identified $b = t$ and $a = k + 1$. The posterior mean is

$$E(\lambda \mid k, t) = \frac{k+1}{t}. \qquad (3.3.4)$$

Posterior density for exchangeable items in n component positions. Consider the case of n component positions using exchangeable items. If the system operates for time t with a total of k observed failures over all n component positions, the likelihood would be

$$P[N(t) = k \mid \lambda] = \frac{(nt\lambda)^k e^{-nt\lambda}}{k!}$$

with MLE

$$\hat{\lambda} = \frac{k}{nt}.$$

If we use a "flat prior," $p(\lambda) = $ constant, the posterior density for λ would be

$$p(\lambda \mid n, k, t) = \frac{(nt)^{k+1} \lambda^k e^{-nt\lambda}}{k!} \qquad (3.3.5)$$

with posterior mean

$$E(\lambda \mid n, k, t) = \frac{k+1}{nt}.$$

Using a spreadsheet such as EXCEL, it is easy to plot posterior densities corresponding to failure observations.

Example 3.3.1 (*Continued*). The normal approximation to the gamma.
In Example 3.3.1, $p(\lambda_A \mid k = 2500, T = 250M)$ is gamma with shape parameter
$a = 2501$. In this example, $T = 250M$ is the total of all hard disk seeks in the
field.

It is neither practical nor desirable to graph a gamma density with a shape
parameter so large. In this case, the mean based on the gamma distribution is
$\mu_A = \frac{2501}{250M} \cong 10^{-5}$, while the standard deviation is $\sigma_A = \frac{\sqrt{2501}}{250M} = 0.025 \times 10^{-5}$.
From Theorem 2.2.2 we know that the sum of iid exponential random quantities
has a gamma distribution. Hence with $a = 2501$ and $\sigma_A = 0.025 \times 10^{-5}$ we can
appeal to the central limit theorem to approximate the gamma density by the
normal density with the same mean and standard deviation as the gamma.

Using the normal approximation, we calculate that the failure rate for Company A satisfies

$$0.95 \times 10^{-5} = \mu_A - 2\sigma_a \le \lambda_A \le \mu_A + 2\sigma_A = 1.05 \times 10^{-5} \qquad (3.3.6)$$

with probability greater than 95%.

The normal approximation for λ_B is still good in this case even though we
now have only $a = 6$. For λ_B we have

$$\mu_B = 0.5 \times 10^{-5} \quad \text{and} \quad \sigma_B = 0.223 \times 10^{-5},$$
$$0.056 \times 10^{-5} = \mu_B - 2\sigma_B \le \lambda_B \le \mu_B + 2\sigma_B = 0.946 \times 10^{-5}. \qquad (3.3.7)$$

Comparing (3.3.6) and (3.3.7) we see that there is virtually no overlap in the
posterior probability densities. It follows that the failure rate for hard disks
from Company B is significantly better than the failure rate for hard disks from
Company A.

Figure 3.3.2 demonstrates graphically the differences in the failure rates of
Companies A and B.

Exercises

Compute all posterior densities using the "flat" prior for λ.
3.3.1. Using the data in Table 3.3.2, compute and graph the posterior densities for fuse and capacitor failure rates. For fuses, compute and graph

$$P[N(t_0) = j \mid \text{fuse data}]$$

for $j = 0, 1, \ldots, 10$ and $t_0 = 1000$ shots. The fuse data consists of $k = 6$ failures
in $t = 700$ shots for $n = 288$ fuses.
3.3.2. In Example 3.3.1 we saw that the posterior densities for λ_A and λ_B
will hardly overlap. However, it would be useful to compute the probability that
failure rates, Λ_A, Λ_B considered as random quantities satisfy $\Lambda_A \le \Lambda_B$.

Consider the random quantities conditionally independent given the data in
Table 3.3.1 and compute $P(\Lambda_A - \Lambda_B \le 0 \mid \text{Data})$ using the normal approximation.
3.3.3. For Company A and a single hard drive, compute and graph

$$P[N(t_0) = j \mid \text{Company A data}]$$

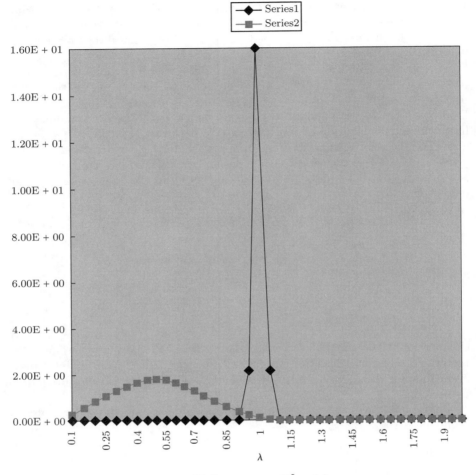

(Failure rate per 10^5 seeks)

FIG. 3.3.2. *Company A: Series* 1. *Company B: Series* 2.

for $j = 0, 1, \ldots, 20$ and $t_0 = 10^6$ seeks. (*Hint.* Using the negative binomial distribution.)

3.4*. Derivation of the Poisson process.

In this section, the population size is again considered conceptually infinite, although the time interval τ and the time horizon T may be finite. In practice, the time horizon T of interest is always finite. Also the time interval τ in which failures are recorded cannot be arbitrarily small. When both $\tau \downarrow 0$ and $T \uparrow \infty$ we obtain the Poisson process model. Although the Poisson process is a very convenient mathematical approximation for analyzing numbers of failures in time intervals, there are situations where other models might be more appropriate.

We will derive the Poisson process starting with finite time interval and finite time horizon models to show the assumptions underlying the Poisson process.

A random process (or stochastic process) is a collection of random quantities, say, $\{X(t_1), X(t_2), \ldots, X(t_n)\}$, at time points (or epochs) $t_1 < t_2 < \cdots < t_n$. In our modeling leading to the Poisson process, time intervals will originally have length τ; e.g.,

where $T = K\tau$ is the time horizon and $\theta(T)$ is the total number of failures in the time interval $[0, T]$.

Let $Y_i = X(\tau_i) - X(\tau_{i-1})$ be the number of failures in the ith time interval $[t_{i-1}, t_i)$ where $t_i = i\tau$. In our derivation of the Poisson process, we consider four related models depending on the finiteness of T and whether or not $\tau \downarrow 0$.

In the first case with both $\tau > 0$ and T finite, we derive the finite population version of the geometric discrete distribution. This is the discrete version of the finite population joint exponential density presented in Chapter 1. The parameter $\theta(T)$ is now the total number of failures in the time interval $[0, T]$. Marginal random variables, $Y_1, Y_2, \ldots, Y_K$, are dependent given $\theta(T)$.

In the second case, we keep the length of an interval $\tau > 0$ but allow the time horizon T to become infinite. In this fashion we derive the usual form of the geometric discrete distribution. This is the discrete version of the infinite population joint exponential density derived in Chapter 1. Note that when $T \uparrow \infty$, $Y_1, Y_2, \ldots, Y_m$ are now *conditionally independent* given the parameter $\lambda = \frac{\theta(T)}{T}$.

In the third case, the time horizon T is kept finite and fixed while the time interval τ is allowed to approach zero. In this way we drive the multinomial discrete distribution with fixed parameter $\theta(T)$. Fix $0 < t_1 < \cdots < t_m = T$ and let $n(t_i)$ be the number of failures in $[t_{i-1}, t_i)$. Random quantities $n(t_1), n(t_2), \ldots, n(t_m)$ are still dependent given $\theta(T)$ as in the first case. However, now we have a property called *orderliness*. By orderliness we mean that two or more failures will *not occur* in the same "infinitesimal" time interval with probability one.

In the fourth and final case, we have the Poisson process where the time horizon has become infinite and the time interval τ is allowed to approach zero. For Poisson processes, we have (1) independent increments $Y(t_i) = X(t_i) - X(t_{i-1})$, (2) orderliness, and (3) marginal random variables have the Poisson distribution; i.e., $P[Y(t_i) = k \mid \lambda] = \frac{(\lambda(t_i - t_{i-1}))^k e^{-\lambda(t_i - t_{i-1})}}{k!}$.

I. The finite population version of the geometric distribution. To obtain the finite population version of the geometric model we again make a stronger judgment than exchangeability. In the following derivation, K in Table 3.4.1 is replaced by N, while $\theta(T)$ in Table 3.4.1 is replaced by S following our notation in section 3.1.

TABLE 3.4.1
Derivation of the Poisson process.

	$T < \infty$	$T \to \infty$
	I	II
$\tau > 0$	Finite population version of the geometric	Independent geometrics (independent discrete increments)
	$\theta(T) < \infty$	$\lim\limits_{\tau \to \infty} \frac{\theta(T)}{T} = \lambda$
	III	IV
$\tau \downarrow 0$	Multinomial (orderliness)	Poisson process (independent increments and orderliness)
	$\theta(T) < \infty$ $K = T/\tau \uparrow \infty$	

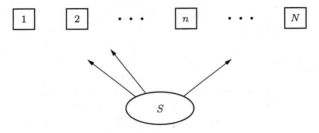

FIG. 3.4.1. *Allocating defects to boxes.*

Imagine N boxes and S nondistinguishable defects. S can be greater than N. (The boxes are our intervals with $\tau = 1$.) See Figure 3.4.1. Suppose now that defects are allocated to boxes in such a way that any box can have $0, 1, \ldots, S$ defects subject to the total of defects in all boxes being S. Let y_i be the number of defects in box i. Suppose we are indifferent regarding all possible vectors $\mathbf{y}_N = (y_1, y_2, \ldots, y_N)$ with the same sum; i.e.,

$$p(\mathbf{y}_N) = p(\mathbf{y'_N}) \quad \text{when} \quad \sum_{i=1}^{N} y_i = \sum_{i=1}^{N} y'_i.$$

The conditional probability function is determined and is

$$p(\mathbf{y}_N \mid N, S) = \frac{1}{\binom{N+S-1}{S}}. \tag{3.4.1}$$

To prove this we need only prove that there are exactly $\binom{N+S-1}{S}$ vectors whose sum of coordinates equals S. Denote the S defects by stars and indicate the N partitioning (or cells) by the N spaces between $N + 1$ bars. Thus

$$|**|*|||*****|$$

is used as a symbol for a distribution of $S = 8$ defects in $N = 5$ cells with occupancy numbers

$$2, 1, 0, 0, 5.$$

Such a symbol necessarily starts and ends with a bar, but the remaining $N - 1$ bars and S stars can appear in an arbitrary order. In this way it becomes apparent that the number of distinguishable vectors equals the number of ways of selecting S places out of $N + S - 1$, namely,

$$\binom{N + S - 1}{S}.$$

Using the same type of argument we can show that

$$p(\mathbf{y}_n \mid N, n, S) = \frac{\binom{N-n+S-s-1}{S-s}}{\binom{N+S-1}{S}}, \tag{3.4.2}$$

where $\sum_{i=1}^{n} y_i = s$. Is $p(\mathbf{y}_n \mid N, n, S)$ exchangeable?
For $n = 1$,

$$\begin{aligned} p(y_1 \mid N, S) &= \frac{\binom{N+S-y_1-2}{S-y_1}}{\binom{N+S-1}{S}} \\ &= \frac{(N + S - y_1 - 2)! S! (N - 1)!}{(S - y_1)!(N - 2)!(N + S - 1)!}. \end{aligned} \tag{3.4.3}$$

In general, fixing n and letting N become infinite in such a way that $\frac{S}{N} \to \lambda > 0$, we have

$$\lim_{N \uparrow \infty, \frac{S}{N} \to \lambda} p(\mathbf{y}_n \mid N, n, S) = \prod_{i=1}^{n} \left[\left(\frac{\lambda}{1 + \lambda} \right)^{y_i} \left(\frac{1}{1 + \lambda} \right) \right] \tag{3.4.4}$$

for $\mathbf{y}_n = (y_1, y_2, \ldots, y_n)$.

In the limit (as $N \to \infty$) random quantities $Y_1, Y_2, \ldots, Y_n, \ldots$ are conditionally independent with univariate geometric probability function

$$p(y_1 \mid \lambda) = \left(\frac{\lambda}{1 + \lambda} \right)^{y_1} \frac{1}{1 + \lambda}$$

for $y_1 = 0, 1, \ldots$. In this case $E(Y_1) = \lambda$, since $E[Y_1 + Y_2 + \cdots + Y_N \mid \sum_{i=1}^{N} Y_i = S] = S$ implies by exchangeability that $E[Y_i \mid \sum_{i=1}^{n} Y_i = S] = \frac{S}{N}$. Hence

$$\lim_{N \uparrow \infty, \frac{S}{N} \to \lambda} E \left[Y_1 \mid \sum_{i=1}^{N} Y_i = S \right] = \lambda.$$

Remark. Note that exchangeability alone does not imply that all vectors with the same sum have constant probability.

Letting $s = \sum_{i=1}^{n} y_i$ and using the same argument as above we can show that

$$p(s \mid N, n, S) = \frac{\binom{n+s-1}{s}\binom{N-n+S-s-1}{S-s}}{\binom{N+S-1}{S}} \qquad (3.4.5)$$

for $0 \leq s \leq S$. Fixing n and taking the limit as $N \to \infty$ and $\frac{S}{N} \to \lambda$ we obtain the negative binomial probability function

$$\lim_{N \to \infty} p(s \mid S = N\lambda) = \binom{n+s-1}{s} \rho^s (1-\rho)^n \qquad (3.4.6)$$

for $s = 0, 1, 2, \ldots$, where $\rho = \frac{\lambda}{1+\lambda}$.

I → II. Independent geometric distributed random quantities.

We assume that $p(\mathbf{y}_k)$ is constant for $\{\mathbf{y}_K \mid = \sum_{i=1}^{K} y_i = \theta(T)\}$ and for every $T > 0$. To go from I to II in Table 3.4.1 we let $T \uparrow \infty$ while at the same time keeping

$$\frac{\theta(T)}{T} = \lambda,$$

where λ is interpreted as the average number of events occurring per unit time. In general, fixing k and letting $K = \frac{T}{\tau}$ become infinite in such a way that $\frac{\theta(T)}{T} = \lambda$, we have

$$\lim_{T \to \infty} p\left(\mathbf{y}_k \mid \frac{\theta(T)}{T} = \lambda\right) = \prod_{i=1}^{k} \left(\frac{\lambda\tau}{1+\lambda\tau}\right)^{y_i} \frac{1}{1+\lambda\tau}, \qquad (3.4.7)$$

where $y_i = 0, 1, \ldots$. It follows from (3.4.7) that $Y_1, Y_2, \ldots, Y_k$ are conditionally independent in the limit as $T \uparrow \infty$ given λ. Also from (3.4.7) it follows that $E(Y_1 \mid \lambda, \tau) = \lambda\tau$.

II → IV. First derivation of the Poisson process.

We now assume that the original $p(\mathbf{y}_K)$ is constant for $\{\mathbf{y}_K \mid \sum_{i=1}^{K} y_i = \theta(T)\}$ and for every T and $\tau > 0$. Consider fixed time points

$$t_1 < t_2 < \cdots < t_m.$$

Let $n(t_i)$ be the number of failures in $[t_{i-1}, t_i)$, and for mathematical convenience assume that t_i's coincide with multiples of the τ's so that

$$t_i = k_i \tau.$$

Then $n(t_i)$ is approximately the sum of $k_i - k_{i-1}$ independent geometric distributed random quantities by II. We need only show that for the fixed interval $[t_{i-1}, t_i)$

$$\lim_{\tau \downarrow 0} P[n(t_i) = r \mid \lambda] = \frac{[\lambda(t_i - t_{i-1})]^r e^{-\lambda(t_i - t_{i-1})}}{r!}. \qquad (3.4.8)$$

Since $t_i - t_{i-1} = (k_i - k_{i-1})\tau$ it follows that in taking the limit, $k_i - k_{i-1}$ must go to infinity as $\tau \downarrow 0$.

To show (3.4.8) we use generating functions. The generating function corresponding to Y_1, the number of events occurring in $[0, \tau)$, is

$$G(z) = E[z^{Y_1}] = \sum_{y_1=0}^{\infty} z^{y_1} \left(\frac{\lambda\tau}{1+\lambda\tau} \right)^{y_1} \left(\frac{1}{1+\lambda\tau} \right) = \left(\frac{1}{1+\lambda\tau(1-z)} \right).$$

The generating function corresponding to $n(t_1) = \sum_{j=1}^{k_1} y_j$, the number of events which occur in $[0, t_1)$, is

$$[G(z)]^{k_1} = \left(\frac{1}{1+\lambda\tau(1-z)} \right)^{k_1}$$

$$= \left(\frac{1}{1+\lambda\frac{t_1}{k_1}(1-z)} \right)^{k_1}$$

since $t_1 = k_1\tau$ and $t_0 = 0$. Letting $k_1 \uparrow \infty$ as $\tau \downarrow 0$, we have

$$\lim_{k_1 \uparrow \infty} [G(z)]^{k_1} = e^{\lambda t_1(z-1)},$$

which is the generating function of a Poisson random quantity with parameter λt_1; that is,

$$P[n(t_1) = r \mid \lambda t_1] = \frac{[\lambda t_1]^r e^{-\lambda t_1}}{r!}$$

as was to be shown.

Independence of $n(t_1), n(t_2), \ldots, n(t_m)$ is true for each $\tau > 0$ where $t_i = k_i\tau > 0$ in case II and therefore true in the limit as $\tau \downarrow 0$. □

I → III. Derivation of the multinomial. Again we consider fixed

$$0 \equiv t_0 < t_1 < t_2 < \cdots < t_m = T,$$

where these coincide with multiples of the τ's. Let $n(t_i)$ be the number of events that occur in $[t_{i-1}, t_i)$. Since the t_i's are integer multiples of the τ's,

$$t_i - t_{i-1} = r_i\tau$$

for some integer r_i. Let $t_m = K\tau = T$.

We claim that in the limit, as $\tau \downarrow 0$, the joint probability function for $n(t_1), n(t_2), \ldots, n(t_m)$ is the multinomial; i.e.,

$$\lim_{\tau \downarrow 0} p[n(t_1), \ldots, n(t_m) \mid \theta(T), T]$$

$$= \left(\begin{array}{c} \theta(T) \\ n(t_1), \ldots, n(t_m) \end{array} \right) \prod_{i=1}^{m} \left(\frac{t_i - t_{i-1}}{T} \right)^{n(t_i)},$$

where $\sum_{i=1}^{m} n(t_i) = \theta(T)$ and $\sum_{i=1}^{m}(t_i - t_{i-1}) = T$.

We will only give the result for $m = 2$ since the general result follows along similar lines. Fix $t > T$. We show that in the limit, as $\tau \downarrow 0$, $n(t)$ given $\theta(t)$ and T is binomial with parameters $\theta(T)$ and $\frac{t}{T}$.

Let $\sum_{i=1}^{k} y_i = n(t)$ where $t = k\tau$. In our discussion of the finite population version of the geometric it was pointed out that

$$P\left[\sum_{i=1}^{k} y_i = n(t) \mid \theta(T), T\right] = \frac{\binom{k+n(t)-1}{n(t)}\binom{K-k+\theta(T)-n(t)-1}{\theta(T)-n(t)}}{\binom{K+\theta(T)-1}{\theta(T)}}$$

for $0 \le n(t) \le \theta(T)$. The right-hand side can be written as

$$\binom{\theta(T)}{n(t)} \frac{\frac{(k+n(t)-1)!(K-k+\theta(T)-n(t)-1)!}{(k-1)!(K-k-1)!}}{\frac{(K+\theta(T)-1)!}{(K-1)!}}.$$

Since $k = \frac{t}{\tau}$ and $K = \frac{T}{\tau}$, we have

$$\binom{\theta(T)}{n(t)} \frac{[(\frac{t}{\tau}+n(t)-1)\cdots(\frac{t}{\tau})][(\frac{T}{\tau}-\frac{t}{\tau}+\theta(T)-n(t)-1)\cdots(\frac{T}{\tau}-\frac{t}{\tau})]}{(\frac{T}{\tau}+\theta(T)-1)\cdots(\frac{T}{\tau})}.$$

In the numerator, the first factor has $n(t)$ terms while the second factor has $\theta(T) - n(t)$ terms. The denominator has $\theta(T)$ terms. Multiply numerator and denominator by τ and let $\tau \downarrow 0$. We have

$$\lim_{\tau \downarrow 0} p[n(t) \mid \theta(T), T]$$

$$= \binom{\theta(T)}{n(t)} \left(\frac{t}{T}\right)^{n(t)} \left[1 - \frac{t}{T}\right]^{\theta(T)-n(t)}.$$

Note that as $t \downarrow 0$, $n(t)$ tends to be either 0 or 1 with high probability. This corresponds to the property that we have called orderliness. □

III → IV. The second derivation of the Poisson process.

We can derive the Poisson probability function from the multinomial. To see how that approach would work, fix m and $t_1 < t_2 < \cdots < t_m < t_m < T$. Since we are now going to let $t_{m+1} = T \uparrow \infty$ while $\frac{\theta(T)}{T} = \lambda$, consider an additional interval $[t_m, T)$ so that

$$t_1 < t_2 < \cdots < t_m < t_{m+1} = T.$$

As before, let $n(t_i)$ be the number of events in $t_i - t_{i-1}$. We next let $t_{m+1} = T \uparrow \infty$ while keeping

$$\sum_{i=1}^{m+1} n(t_i) = \theta(T) = T\lambda.$$

Let $\sum_{i=1}^{m} n(t_i) = s$ so that

$$p\left(n(t_1), \ldots, n(t_m) \mid \sum_{i=1}^{m+1} n(t_i) = \theta(T) = T\lambda\right)$$

$$= \binom{s}{n(t_1), \ldots, n(t_m)} \prod_{i=1}^{m} \left(\frac{t_i - t_{i-1}}{t_m}\right)^{n(t_i)} \binom{T\lambda}{s} \left(\frac{t_m}{T}\right)^s \left(1 - \frac{t_m}{T}\right)^{T\lambda - s}.$$

Letting $T \uparrow \infty$ while $\frac{\theta(T)}{T} = \lambda$, we obtain

$$\lim_{T \uparrow \infty} p\left(n(t_1), \ldots, n(t_m) \mid \sum_{i=1}^{m+1} n(t_i) = \theta(T) = T\lambda\right)$$

$$= \prod_{i=1}^{m} \frac{[\lambda(t_i - t_{i-1})]^{n(t_i)} e^{-\lambda(t_i - t_{i-1})}}{n(t_i)!},$$

which is the joint probability function for independent Poisson random quantities.

Exercises

3.4.1. For the finite population exponential model

$$p\left(x_1, x_2, \ldots, x_n \mid x_i \geq 0, \sum_{i=1}^{N} x_i = N\theta\right)$$

$$= \frac{(N-1)}{N\theta} \frac{(N-2)}{N\theta} \cdots \frac{(N-n)}{N\theta} \left[1 - \frac{\sum_{i=1}^{n} x_i}{N\theta}\right]^{N-n-1},$$

calculate $P[N(t) = k \mid \sum_{i=1}^{N} x_i = N\theta]$ for $0 \leq k < N$. Compare this expression with the corresponding infinite population exponential model which gives the Poisson probability

$$P[N(t) = k \mid \theta] = \frac{\left(\frac{t}{\theta}\right)^k}{k!} e^{-\frac{t}{\theta}}.$$

3.5.   Notes and references.

Section 3.1. The material in this section is related to the material in section 3.2. In particular, the multinomial in section 3.2 is the generalization of the binomial considered here, while the Dirichlet is the generalization of the beta. The material on failure diagnosis is similar to the medical example in the paper by Pereira (1990).

Section 3.2. For generalizations of the example concerning missing data in this section, see Basu and Pereira (1982).

Section 3.3. Examples 3.3.1 and 3.3.2 are related to problems encountered in practice.

Section 3.4. This section was motivated by the unpublished work of Hesselager, Mendel, and Shortle. The models considered here are also useful for predicting the number of arrivals to a server.

Notation

Random quantity	Probability name	Probability function
λ	gamma	$g_{a,b}(\lambda) = \dfrac{b^a \lambda^{a-1} e^{-b\lambda}}{\Gamma(a)} \quad \lambda \geq 0, a, b > 0$
	cumulative	$G_{a,b}(\lambda) = P_{a,b}(\Lambda \leq \lambda) \quad a$ integer
		$= 1 - \displaystyle\sum_{i=0}^{a-1} \dfrac{(b\lambda)^i}{i!} e^{-b\lambda}$
$N(t)$	cumulative number of failures in $[0, t]$	
$N(t) \mid \lambda$	Poisson	$P[N(t) = k \mid \lambda] = \dfrac{(\lambda t)^k e^{-\lambda t}}{k!} \quad k = 0, 1, 2, \ldots$
$N(t)$	unconditional when λ	
	is gamma (a, b)	$\displaystyle\int_0^\infty P[N(t) = k \mid \lambda] g_{a,b}(\lambda) d\lambda$
		$= \dfrac{\Gamma(a+k)}{\Gamma(k+1)\Gamma(a)} \left(\dfrac{b}{b+t}\right)^a \left(\dfrac{t}{b+t}\right)^k \quad k = 0, 1, 2, \ldots$
	(negative binomial)	
$s \mid N, n, S$	hypergeometric	$p(s \mid S, n, N) = \dfrac{\left(\begin{array}{c} S \\ s \end{array}\right)\left(\begin{array}{c} N - S \\ n - s \end{array}\right)}{\left(\begin{array}{c} N \\ n \end{array}\right)}$
		$s = 0, 1, \ldots, \min(n, S)$
$s \mid n, \rho$	binomial	$p(s \mid n, \rho) = \left(\begin{array}{c} n \\ s \end{array}\right) \rho^s (1 - \rho)^{n-s} \quad s = 0, 1, \ldots, n$
$\rho \mid A, B$	beta	$\dfrac{\Gamma(A+B)}{\Gamma(A)\Gamma(B)} \rho^{A-1} (1 - \rho)^{B-1} \quad A, B > 0, \quad 0 \leq \rho \leq 1$
	$Be(A, B)$	
$S \mid N, A, B$	beta-binomial	$\displaystyle\int_0^1 \left(\begin{array}{c} N \\ S \end{array}\right) \rho^S (1 - \rho)^{N-S} \dfrac{\Gamma(A+B)}{\Gamma(A)\Gamma(B)} \rho^{A-1}$
		$\cdot (1 - \rho)^{B-1} d\rho$
	$Bb[N,A,B]$	

CHAPTER 4

Strength of Materials

4.1. Introduction.

There are two basic kinds of failure:

    (1) *wearout*

and

    (2) *overstress.*

Wearout implies that a device or material becomes unusable through long or heavy use. It implies the using up or gradual consuming of material. We have already discussed the analysis of wearout data. The jet engine data analyzed in Chapter 2 are an example of the occurrence of wearout.

Overstress, on the other hand, refers to the event that an applied stress exceeds the strength of a device or material. This is the type of failure examined in this chapter. The problem is to determine whether or not a specified or random load will exceed the strength of a device or material. Both the strength and the load may be random. However, they are usually considered independent of one another. In section 4.2 we consider the problem of assigning a probability distribution to the strength of a material based on test data. In section 4.3 we discuss design reliability. That is, based on projected principal stresses, we calculate the maximum shear stress that could result.

According to classical theory, the ultimate strength of a material is determined by the internal stresses at a point. We begin with a simple illustrative example, called a uniaxial tensile test. A coupon of material is stressed by pulling on it with an applied force until it yields (Figure 4.1.1). The applied force required to cause yielding is equivalent to the internal strength of the material. By repeating this experiment for a given type of material, aluminum, for example, we will obtain material strength data. However, experimental measurements will give many results that may hardly agree with any deterministic theory. For this reason, we require a probability distribution to predict strength.

How can we predict failure? We will consider two criteria relevant to this prediction:

    (1) the *distortion energy* criterion due to von Mises

and

    (2) the maximum shear stress criterion. This will be discussed in section 4.3.

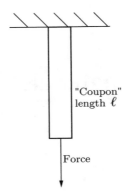

FIG. 4.1.1. *Uniaxial tensile test.*

The distortion energy criterion is useful in relating uniaxial yield strength data to more realistic complex loading conditions. Also, test specimens are typically much smaller than the actual structural members. Hence the distortion energy criterion is useful for scaling test results up to full-scale members.

In the isotropic linear theory of elasticity we can decompose the total elastic strain energy *per unit volume of material* as follows:

$$\text{total elastic energy} = \text{distortion energy} + \text{volumetric energy.} \qquad (4.1.1)$$

Volumetric energy is the energy stored in a material due to a pure volume change. For example, were we to submerge a cube of material at some depth in the ocean, the symmetric effect of hydrostatic pressure on the cube would result in a pure volume change.

According to the von Mises–Hencky criterion, when the distortion energy of a material reaches a certain value, the material will yield. Since we are interested in material failure, the distortion energy component of total elastic energy is important to us. This is the only energy component relevant to a tensile test. A basic reference is *The Feynman Lectures on Physics*, Vol. II, Chapter 38.

Hooke's law.   Imagine pulling the coupon in Figure 4.1.1 until yielding occurs. The stress imposed is by definition the force imposed per unit area. Although we have given a two-dimensional picture of our coupon, it is really three dimensional, and the area in question is not shown. As we pull the coupon, it elongates and at the same time the widths contract. As the experiment proceeds, we obtain the graph in Figure 4.1.2. This is called Hooke's law and is based on empirical observations which indicate a linear relationship between stress σ and strain $e = \frac{\Delta \ell}{\ell}$ at the moment of yielding, i.e., *Hooke's law*

$$\text{force} = EA \frac{\Delta \ell}{\ell}, \qquad (4.1.2)$$

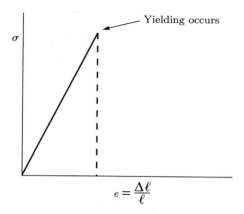

FIG. 4.1.2. *Stress versus strain.*

where A is the bottom area of the coupon and E is the modulus of elasticity, a material constant. Since stress σ by definition is the force per unit area, we have

$$\sigma = E\frac{\Delta\ell}{\ell} = Ee.$$

The work done to effect yielding is equal to the total elastic energy corresponding to strength. This can be calculated as the area under the line in the graph to the point of yielding or

$$\text{total elastic energy} = \frac{\sigma e}{2} = \frac{\sigma^2}{2E}, \qquad (4.1.3)$$

where E is Young's modulus or the modulus of elasticity. Since we are interested in failure due to yielding, we want to calculate the distortion energy. We show in section 4.3 that in this case

$$\text{distortion energy} = \frac{\sigma^2}{6G}, \qquad (4.1.4)$$

where $G = \frac{E}{2(1+v)}$, called the shear modulus, is related to both Young's modulus E and Poisson's ratio v.

When you stretch a block of material, it also contracts at right angles to the stretch. Poisson's ratio relates these contractions to the strain $e = \frac{\Delta\ell}{\ell}$. Namely, if w stands for width, h stands for height, and ℓ stands for length, then

$$\frac{\Delta w}{w} = \frac{\Delta h}{h} = -v\frac{\Delta\ell}{\ell}, \qquad (4.1.5)$$

where v is Poisson's ratio. The constants E and v completely specify the elastic properties of a homogeneous isotropic (noncrystalline) material. (See Figure 4.1.3.)

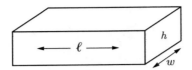

FIG. 4.1.3. *Block of material.*

Exercises

4.1.1. Calculate the total elastic energy in terms of σ when instead of Hooke's law we have

$$\sigma \propto \left(\frac{\Delta\ell}{\ell}\right)^m,$$

where $m > 0$.

4.2. Strength of materials and the Weibull distribution.

In 1939, a Swedish mechanical engineer by the name of Walodie Weibull published a very influential paper, "A Statistical Theory of the Strength of Materials." In this paper he notes that the classical mechanical engineering theory of static failure is deterministic and is insufficient to explain experiment data. He "derived" a probability distribution for strength subsequently called the Weibull distribution. His "derivation" has been criticized in a paper by Lindquist (1994), and in what follows we follow Lindquist's derivation.

Consider N exchangeable items and the distortion energy required to cause yielding. For item i

$$u_i = \frac{\sigma_i^2}{6G}, \tag{4.2.1}$$

$i = 1, 2, \ldots, N$ is the distortion energy relative to yielding. As in Chapter 1, N is the finite population size. Let $(u_1, u_2, \ldots, u_N)$ be the vector of possible distortion energies for the N items. Suppose, furthermore, that we would be indifferent between $(u_1, u_2, \ldots, u_N)$ and $(u'_1, u'_2, \ldots, u'_N)$ if

$$\sum_{i=1}^{N} u_i = \sum_{i=1}^{N} u'_i$$

in the sense that in our judgment

$$p(u_1, u_2, \ldots, u_N) = p(u'_1, u'_2, \ldots, u'_N).$$

It follows from Theorem 1.1.1 that

$$p\left(u_1, u_2, \ldots, u_n \mid \sum_{i=1}^{N} u_i = N\theta\right)$$

$$= \frac{(N-1)}{N\theta} \frac{(N-2)}{N\theta} \cdots \frac{(N-n)}{N\theta} \left[1 - \frac{\sum_{i=1}^{n} u_i}{N\theta}\right]^{N-n-1}$$

and by equation (1.3.1),

$$P\left(U_1 > y_1, U_2 > y_2, \ldots, U_n > y_n \mid \sum_{i=1}^{N} U_i = N\theta\right)$$

$$= \left[1 - \sum_{i=1}^{n} \frac{y_i}{N\theta}\right]^{N-1}.$$

Let X_i be the random stress leading to yielding and $U_i = \frac{X_i^2}{6G} = \psi(X_i)$ for item i so that

$$P\left(X_1 > \sigma_1, X_2 > \sigma_2, \ldots, X_n > \sigma_n \mid \sum_{i=1}^{N} U_i = N\theta\right)$$

$$= P\left(\psi(X_1) > \psi(\sigma_1), \ldots, \psi(X_n) > \psi(\sigma_n) \mid \sum_{i=1}^{N} U_i = N\theta\right) \qquad (4.2.2)$$

since ψ is strictly increasing. Since $U_i = \psi(X_i)$ we have

$$P\left(U_1 > \psi(\sigma_1), \ldots, U_n > \psi(\sigma_n) \mid \sum_{i=1}^{N} U_i = N\theta\right)$$

$$= \left[1 - \sum_{i=1}^{n} \frac{\psi(\sigma_i)}{N\theta}\right]^{N-1}.$$

Substituting $\frac{\sigma_i^2}{6G}$ for $\psi(\sigma_i)$ and taking the limit as $N \uparrow \infty$ we have

$$\lim_{N\uparrow\infty} P\left(X_1 > \sigma_1, \ldots, X_n > \sigma_n \mid \sum_{i=1}^{N} U_i = N\theta\right)$$

$$= \lim_{N\uparrow\infty} P\left(U_1 > \frac{\sigma_i^2}{6G}, \ldots, U_n > \frac{\sigma_n^2}{6G} \mid \sum_{i=1}^{N} U_i = N\theta\right)$$

$$= \lim_{N\uparrow\infty} \left[1 - \sum_{i=1}^{N} \frac{\sigma_i^2}{6GN\theta}\right]^{N-1} = \prod_{i=1}^{n} e^{-\frac{\sigma_i^2}{6G\theta}}.$$

Hence

$$P(X_i > \sigma_i \mid \theta) \overset{DEF}{=} \overline{F}(\sigma_i \mid \theta) = e^{-\frac{\sigma_i^2}{6G\theta}} \qquad (4.2.3)$$

is the Weibull survival distribution with shape parameter $\alpha = 2$!

The Weibull shape parameter is not always 2. It does not follow, however, that a Weibull distribution with shape parameter 2, as in formula (4.2.3), can always be defended as the distribution for strength of a material. For example, suppose we judge the following:

(1) there exists a minimum distortion energy u_0, which any specimen can absorb before yielding, and

(2) the material is no longer linear elastic. The distortion energy for the ith specimen in uniform tension (or compression) is for this case represented as a more general function

$$U(\sigma_i) = K\sigma_i^m$$

for $\sigma_i > \sigma_0$, where $u_0 = U(\sigma_0)$ and K and m are positive constants for the material under consideration.

In this case, it can be shown that

$$p(\sigma_i \mid \theta, \sigma_0, m) = \left[\frac{mK}{\theta - \sigma_0^m}\right] \sigma_i^{m-1} e^{-K\frac{(\sigma_i^m - \sigma_0^m)}{(\theta - \sigma_0^m)}} \tag{4.2.4}$$

for $\sigma_i > \sigma_0$, where $\theta = \lim_{N \to \infty} \frac{\sum_{i=1}^{N} U_i}{N}$ as before.

Load application. Let the random quantity S denote the strength of a specified material. Suppose a "coupon" of this material is subjected to uniaxial loading as in Figure 4.1.1. Let the random quantity L denote the maximum loading from the beginning of application until the load is removed. We require

$$P(S > L),$$

i.e., the probability that the coupon does not yield under the random loading. Usually S and L are assumed to be independent random quantities. If

$$P(S > s \mid \theta) = e^{-\frac{s^2}{6G\theta}}$$

as in formula (4.2.3), then we need to calculate

$$\int_{\ell=0}^{\infty} e^{-\frac{\ell^2}{6G\theta}} p(\ell)d\ell, \tag{4.2.5}$$

where $p(\ell)$ is the probability density for the loading. Often $p(\ell)$ is assumed to be a normal probability density, although the loading in this case must be positive.

Exercises

4.2.1. Suppose that instead of formula (4.2.1), the distortion energy based on Hooke's law, we use the formula

$$u(\sigma_i) = K\sigma_i^m, \qquad \sigma_i \geq \sigma_0,$$

where K and m are positive constants for the material in question. Following the method of deriving the Weibull distribution in this section where α turned

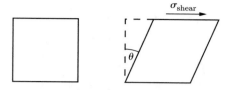

FIG. 4.3.1. *Shear stress.*

out to be 2, show that in this case

$$p(\sigma_i \mid \theta, \sigma_0, m) = \left[\frac{mK}{\theta - \sigma_0^m}\right] \sigma_i^{m-1} e^{-K\frac{(\sigma_i^m - \sigma_0^m)}{(\theta - \sigma_0^m)}} \qquad (4.2.6)$$

for $\sigma_i > \sigma_0$.

4.2.2. The strength S of a certain material is judged to have a Weibull $(\alpha = 2, \theta)$ probability distribution. Suppose a maximum uniaxial load L is applied to a "coupon" of the material as in Figure 4.1.1. If S and L are judged independent and

$$L \sim \text{Normal}(\mu, \sigma^2),$$

where now σ is the standard deviation of the normal distribution and μ is its mean, then develop a computing formula for

$$P(S > L)$$

which can be easily calculated based on the cumulative normal probability distribution.

4.3. Shear stress in biaxial stress.

In this section we consider the effect of *external loads* on a material. Consider the face of a cube of material. Shearing stresses are forces that tend to distort right angles as illustrated in Figure 4.3.1. Again, according to the theory of linear elasticity, stress is equal to a constant times strain or

$$\sigma_{shear} = \frac{E}{2(1 + v)}\theta, \qquad (4.3.1)$$

where $G = \frac{E}{2(1+v)}$ is called the shear modulus.

The maximum shear stress theory assumes that yielding begins when the maximum shear stress in the material becomes equal to the maximum shear stress at the yield point in a simple tension test.

Principal stresses and strains. Consider an infinitesimal cube surrounding a point in an *isotropic* piece of material. Figure 4.3.2 illustrates the stress element in a two-dimensional state of stress. In Figure 4.3.2, the coordinate x, y, z system was chosen arbitrarily. In this figure, σ_x, σ_y are the stresses along coordinate axes while τ_{xy} and τ_{yx} are the equal shear stresses. By appropriate transformation

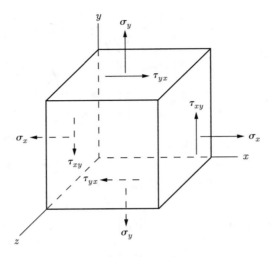

FIG. 4.3.2. *Stress element showing two-dimensional state of stress.*

of the coordinate system we can eliminate the shear stresses, retaining only the *principal stresses* σ_1, σ_2. The two stress-free surfaces of the element are by definition principal planes since they are subject to zero shear stress.

In the tensile test experiment, there was only one nonzero principal stress, namely, σ_1. The other principal stresses $\sigma_2 = \sigma_3 = 0$. In general, these principal stresses are not zero and may be oriented in any direction along the principal orthogonal coordinate axes. Assessing the distribution of the three principal stresses is essential for predicting static failure under multi-axial stress.

According to the von Mises–Hencky criterion for failure, a material fails when the total distortion energy per unit volume due to external loads exceeds the internal distortion energy determined by a tensile test as in the previous section. The total distortion energy per unit volume due to external stresses is

$$\frac{1}{6G} \left(\sigma_1^2 + \sigma_2^2 + \sigma_3^2 - \sigma_1\sigma_2 - \sigma_1\sigma_3 - \sigma_2\sigma_3 \right). \qquad (4.3.2)$$

Setting this equal to the internal distortion energy determined by a tensile test defines an ellipsoid within which external principal stresses can lie without causing failure.

Why 6G in formulas (4.1.1) and (4.3.2) for distortion energy? For the block of material in Figure 4.1.3 let U_e be the total elastic energy, U_{dist} be the distortion energy, and U_{vol} the volumetric energy. Then

$$U_e = U_{dist} + U_{vol}$$

or

$$U_{dist} = U_e - U_{vol} = \frac{\sigma_i^2}{2E} - U_{vol}. \qquad (4.3.3)$$

The volumetric energy can be shown to be

$$U_{vol} = 3 \left(\frac{1}{2} \right) \left(\frac{\sigma_1 + \sigma_2 + \sigma_3}{3} \right) \left(\frac{e_1 + e_2 + e_3}{3} \right),$$

where $\sigma_i, e_i, i = 1, 2, 3$ are the stresses and strains at the moment of yielding.

Suppose $\sigma_2, \sigma_3 \neq 0$; then by the principle of superposition, i.e., when a material is stretched it contracts at right angles to the stretch,

$$e_1 = \frac{\sigma_1}{E} - v\frac{\sigma_2}{E} - v\frac{\sigma_3}{E},$$

$$e_2 = \frac{\sigma_2}{E} - v\frac{\sigma_1}{E} - v\frac{\sigma_3}{E}, \qquad (4.3.4)$$

$$e_3 = \frac{\sigma_3}{E} - v\frac{\sigma_2}{E} - v\frac{\sigma_1}{E}.$$

But in our tensile test $\sigma_2 = \sigma_3 = 0$ so that

$$e_1 = \frac{\sigma_1}{E},$$

$$e_2 = -v\frac{\sigma_1}{E},$$

$$e_3 = -v\frac{\sigma_1}{E}.$$

It follows that

$$U_{dist} = U_e - U_{vol}$$

$$= \frac{1}{2}\frac{\sigma_1^2}{E} - \frac{1}{2E}\sigma_1^2\left(\frac{1-2v}{3}\right)$$

$$= \frac{\sigma_1^2}{2E}\left[\frac{2}{3} + \frac{2v}{3}\right] = \frac{\sigma_1^2}{6}\left[\frac{2(1+v)}{E}\right] \qquad (4.3.5)$$

$$= \frac{\sigma_1^2}{6G},$$

where $G = \frac{E}{2(1+v)}$ is called the *shear modulus*.

Maximum shear stress.   Failure theories generally state that yielding occurs whenever the combination of principal stresses acting at a point exceeds a critical value. The classical von Mises–Hencky criterion, which we used in section 2, postulates that yielding is caused by the distortion energy which is a function of the principal stresses. Another important criterion is the *maximum shear stress* criterion. Maximum shear stress can be found graphically using Mohr's circle.

Mohr's circle.   In the case of biaxial loading, all stresses on a stress element act on only two pairs of faces as shown in Figure 4.3.2. The two stress-free surfaces of the element are by definition principal planes since they are subjected to zero

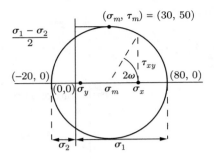

FIG. 4.3.3. *Mohr's circle.*

shear stress. The axis normal to the two stress-free surfaces is a principal axis. The other two principal directions, therefore, are parallel to these stress-free surfaces, and the magnitudes of these two principal stresses are denoted by σ_1 and σ_2 as before. Mohr's circle provides a convenient means for representing the normal and shear stresses acting on planes perpendicular to the stress-free surfaces.

Let σ_x and τ_{xy} denote the normal stress and shear stress acting on the x-plane that is perpendicular to the stress-free surfaces. Given σ_1 and σ_2, Mohr's circle, Figure 4.3.3, corresponds to realizations of σ_x, τ_{xy} as we rotate the x- and y-axes. Let ω be the angle between the positive direction of the x-axis and the first principal stress direction. The rotation angle in Mohr's circle is 2ω. The point on the circle (σ_x, τ_{xy}) corresponds to the state of stress acting on the x-plane.

The state of stress on an infinitesimal cube surrounding a point in a piece of material can be represented by a matrix. In the biaxial case this matrix is

$$\begin{bmatrix} \sigma_x & \tau_{xy} \\ \tau_{xy} & \sigma_y \end{bmatrix}.$$

In the isotropic case, $\tau_{xy} = \tau_{yx}$.

Mohr's circle is another way of representing the state of stress on an infinitesimal cube surrounding a point in a piece of material. The magnitudes of the principal stresses σ_1, σ_2 *and* the orientation angle ω determine the point on the circle (σ_x, τ_{xy}), where

$$\sigma_x = \frac{\sigma_1 + \sigma_2}{2} + \frac{\sigma_1 - \sigma_2}{2} \cos 2\omega,$$

$$\tau_{xy} = \frac{\sigma_1 - \sigma_2}{2} \sin w\omega. \tag{4.3.6}$$

The radius τ_m and the center σ_m of Mohr's circle are determined by

$$\tau_m = \frac{|\sigma_1 - \sigma_2|}{2}, \tag{4.3.7}$$

$$\sigma_m = \frac{\sigma_1 + \sigma_2}{2}. \tag{4.3.8}$$

The maximum shear stress τ_m occurs when $\omega = 45°$. In Figure 4.3.3, $\tau_m = 50$.

To find σ_1 and σ_2 given σ_x, σ_y, and τ_{xy}, use $\sigma_m = \frac{\sigma_1 + \sigma_2}{2} = \frac{\sigma_x + \sigma_y}{2}$. The line segment drawn from $(\sigma_m, 0)$ to (σ_x, τ_{xy}) determines the radius of Mohr's circle τ_m.

Indifference with respect to the x- and y-axis orientation. Conditioning on the magnitudes of the principal stresses σ_1 and σ_2, suppose we are indifferent to the orientation of the x- and y-axes with respect to the principal axes; i.e., we may choose our x- and y-axes in an arbitrary manner. Since the angle ω is not specified, σ_x can only be determined probabilistically. This indifference with respect to ω corresponds to the invariant probability measure on Mohr's circle

$$\frac{1}{2\pi} d\omega,$$

which is invariant under rotations. Differentiating (4.3.6) with respect to ω we obtain

$$d\sigma_x = -(\sigma_1 - \sigma_2)\sin 2\omega \, d\omega$$

or

$$d\sigma_x = \pm 2\sqrt{\tau_m^2 - (\sigma_x - \sigma_m)^2} \, d\omega.$$

Changing to the coordinate σ_x, the invariant probability measure on the circle becomes

$$\frac{1}{2\pi} d\omega = \pm \frac{1}{4\pi} \left[\tau_m^2 - (\sigma_x - \sigma_m)^2\right]^{-\frac{1}{2}} d\sigma_x$$

for $\min(\sigma_1, \sigma_2) < \sigma_x < \max(\sigma_1, \sigma_2)$.

Since there are four possible orientations of the x- and y-axes that will result in observing the same normal stress σ_x and shear stress τ_{xy} (see Figure 4.3.4 below), the desired conditional probability is thus four times the invariant measure above; i.e.,

$$p(\sigma_x \mid \sigma_1, \sigma_2) = \frac{1}{\pi} \left[\tau_m^2 - (\sigma_x - \sigma_m)^2\right]^{-\frac{1}{2}}$$

for $\min(\sigma_1, \sigma_2) < \sigma_x < \max(\sigma_1, \sigma_2)$.

Exercises

4.3.1. Justify equations (4.3.4) using the principle of superposition.

4.3.2. Given that σ_1, σ_2 and σ_x, τ_{xy} and σ_y are determined, use Mohr's circle to calculate τ_{xy} and σ_y in terms of σ_1, σ_2, and σ_x. Assume $\sigma_3 = 0$.

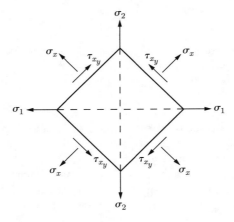

FIG. 4.3.4. *Planes with the same states of stress.*

4.3.3. Calculate $p(\tau_{xy} \mid \sigma_1, \sigma_2)$ assuming indifference with respect to ω; i.e., assume the invariant probability measure on Mohr's circle

$$\frac{1}{2\pi} d\omega.$$

4.4. Probability plots.

In section 4.2 we derived the Weibull distribution starting from Hooke's law. However, it is more common to plot experimental data to "determine" whether or not a Weibull distribution is appropriate. These plots are called "probability plots."

What is a probability plot?   A probability plot is a graph based on the empirical distribution function and relative to a specified family of distributions such as the Weibull, normal, lognormal, etc. The objective is to "fit" a straight line using special probability paper. To illustrate, a Weibull probability plot uses the Weibull survival distribution; i.e.,

$$\overline{F}(x \mid \alpha, \beta) = e^{-\left(\frac{x}{\beta}\right)^{\alpha}}.$$

Taking natural logarithms twice we obtain

$$\ln\ln\left[\frac{1}{\overline{F}(x \mid \alpha, \beta)}\right] = \alpha \ln x - \alpha \ln \beta.$$

Substituting the empirical survival distribution $\overline{F}_n(x)$ for $\overline{F}(x \mid \alpha, \beta)$ we have

$$\ln\ln\left[\frac{1}{\overline{F}_n(x)}\right] = \alpha \ln x - \alpha \ln \beta.$$

The Weibull probability plot is then a plot of

$$\ln \ln \left[\frac{1}{\overline{F}_n(x)} \right] \text{ versus } \ln x.$$

Plotted x's will be the order statistics, i.e., $x = x_{(i)}$, $i = 1, 2, \ldots, n$. Let a straight line be fitted to the plot; call it

$$y = a \ln x + b.$$

The slope of the line, a, "estimates" the shape parameter α, while the scale parameter β is "estimated" by $e^{-b/a}$. The justification for these estimates rests on the hope that

$$\overline{F}_n(x) \cong e^{-\left(\frac{x}{\beta}\right)^\alpha} \tag{*}$$

for some α, β. With exchangeability and *conditional* observations from

$$\overline{F}(x \mid \alpha, \beta) = e^{-\left(\frac{x}{\beta}\right)^\alpha},$$

the limiting result is mathematically true. For a small sample (*) is indeed a stretch!

Engineering textbooks dealing with probability advocate using probability plots; see, e.g., Hahn and Shapiro (1974). These authors feel that even with relatively small sample sizes they can obtain estimates of probability distribution parameters as well as a graphic picture of how well the distribution fits the data. Statistics textbooks written by mathematical statisticians do not mention probability plots. Their reason may be that since estimates are not "true" parameter values, confidence limits need to be determined for the parameters. To obtain confidence limits they use other methods—often likelihood methods.

What is wrong with probability plots?   To determine the relationship between physical variables of interest such as stress and strain in tensile tests, for example, the strain could be measured at various time points as the pulling force is slowly increased. The plot of stress versus strain will never be actually linear since the elasticity assumption is never fully obeyed. This is because a slightly larger amount of work is done in deforming the material than is returned by the material when it relaxes. Hence the linear elasticity assumption is never fully satisfied. However, such plots are useful in determining approximate relationships which can be inferred by drawing a curve or straight line which "fits the data."

This same idea applied to the problem of determining a probability distribution, however, is not valid. In the case of stress and strain, the variables are operationally defined, while this is not the case with probability. Probability as a limiting frequency can never be observed.

For both engineers and many mathematical statisticians, probability is considered to be an "objective" quantity. Furthermore, for engineers and most mathematical statisticians "parameters" usually have *no* operational meaning. In our approach, starting from a finite population, parameters of interest must be functions of observable data. As such they have operational meaning. We feel that only in this case can the analyst have an opinion about parameters. This opinion should then be expressed through a probability distribution. We have tried to use engineering knowledge whenever possible to derive conditional probability models such as the Weibull distribution. In the case of the Weibull distribution, the shape parameter α has no operational meaning in the case of a finite population as does the mean. Hence we advocate using physical approximations such as Hooke's law to determine the shape parameter rather than using data to "estimate" α.

Things to remember. *First*, probability plots are based on a judgment of exchangeability. If, for example, there is a trend in time relative to observations, then exchangeability is not correct. So it is important to first make a "run chart" for observations as a function of the time order in which they are made. This can be useful relative to making the exchangeability judgment, if warranted.

Second, the only justification for probability plots, conditional on exchangeability, is the limiting convergence of the empirical survival distribution $\overline{F}_n$. It is difficult to recommend probability plots for small samples and for very large samples; just use $\overline{F}_n$.

Third, probability plots violate the likelihood principle relative to estimating parameters. That is, the likelihood contains all the information available in the data concerning parameters [D. Basu (1988)].

Fourth, probability plots are based on an eyeball assessment. By taking sufficiently many logarithms we can usually get something that looks like a straight line. Least squares fits also have little or no justification. Probability plots can only be justified as a quick, very preliminary analysis of data.

4.5. Notes and references.

Section 4.1. See Feynman (1964) Chapter 38 for additional discussion of stress and strain. For more discussion of the engineering implications, see Hoffman and Sachs (1953).

Section 4.2. Weibull (1939) derives a cumulative probability distribution of failure of the form

$$F(\sigma) = 1 - e^{-n(\sigma)}$$

and then makes the assumption that

$$n(\sigma) = \frac{(x - x_u)^m}{\sigma_0},$$

where σ is the stress, and x_u, x_0, and m are parameters. The only one of these parameters for which some physical interpretation applies is x_u. The basis for this selection of $n(\sigma)$ is purely empirical since it is determined by curve fitting.

Section 4.3. For further discussion of Mohr's circles see Hoffman and Sachs (1953). The derivation of

$$p(\sigma_x \mid \sigma_1, \sigma_2)$$

is due to Tsai (1994). See also Barlow and Mendel (1992).

For additional discussion of a designer's approach to reliability in mechanical engineering see Vinogradov (1991).

Section 4.4. For additional discussion of probability plots, see Hahn and Shapiro (1974).

Notation

σ	stress, i.e., force per unit area
$e = \dfrac{\Delta \ell}{\ell}$	strain, i.e., fractional change in length
E	Young's modulus
$\dfrac{\sigma^2}{2E}$	total elastic energy based on a tensile test
$\dfrac{\sigma^2}{6G}$	distortion energy based on a tensile test
v	Poisson's ratio
$G = \dfrac{E}{2(1+v)}$	shear modulus
τ_{xy}	shear stress acting on the x-plane in the y-direction

PART II

The Economics of Maintenance Decisions

CHAPTER 5

The Economics of Maintenance and Inspection

5.1. Introduction.

So far we have only concentrated on the uncertainty relative to component and system lifetime. In practice, the cost of inspection and maintenance decisions are often of paramount importance. These too are of course not known with certainty, but it is conventional to use expected costs and compute expected *present worth* or expected *future worth*. First, in section 5.2, we discuss the economic analysis of replacement decisions based on discounted expected costs and present worth but without considering the uncertainty in lifetime.

In Example 5.2.1, the impact of depreciation and other income tax considerations relative to replacement decisions are *not* included as they should be in a realistic comprehensive analysis. Also, the revenues that might be generated by the machines considered are omitted. Due to the time value of money, it is necessary to use a *discount rate* (or interest rate), call it i. Let N be the time horizon under consideration. In our first example $N = 3$.

In section 5.3 we discuss Deming's ALL or NONE rule in inspection sampling. In section 5.4 we study age replacement when lifetimes are uncertain. In section 5.4, we use the expected present worth criterion with respect to continuous discounting; i.e., the interest rate is calculated continuously for an *unbounded* time horizon.

5.2. Deterministic economic analysis.

First let us review discounting. If an amount P in, say, dollars, is put in the bank or invested at a compound interest rate i per time period, then at the end of the time period you will have the future amount $F = P + iP = P(1+i)$ or $P = F(1+i)^{-1}$. In general, if amounts A_i are deposited at the end of time periods $i = 1, 2, \ldots, N$, the present worth of this cash flow at compound interest rate i will be

$$PW(i) = \sum_{t=0}^{N} A_i(1+i)^{-t},$$

where A_0 is the present amount on hand. In deciding when to replace items based on maintenance considerations over more than one year, we should use discounting based on some relevant interest rate.

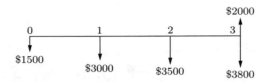

FIG. 5.2.1. *Cash flow if machine is not replaced.*

Example 5.2.1. Discounting over a finite time horizon.

Cost of keeping an old machine. Suppose we are considering the problem of whether to keep an old machine or whether to sell it and buy a new machine. The machine could be a computer, or a truck, etc. The old machine needs an immediate overhaul (time $t = 0$), which we suppose would cost $1500 were we to plan on keeping the old machine. With overhaul, expected maintenance costs for the first year amount to $3000, for the second year $3500, and $3800 for the third year. Maintenance costs are listed for the *end* of each year although they may have been incurred during the year. The present salvage value of the old machine is $10,000, while at the end of the third year we expect that the machine will only have a salvage value of $2000 perhaps due partly to obsolescence. Notice that the expectations are what "we expect." They are not mathematical expectations based on repetitions.

Figure 5.2.1 is a cash flow diagram illustrating the maintenance cost situation. Downward pointing arrows denote costs, while upward pointing arrows indicate income. In our example, we would realize only a $2000 salvage value were we to sell the old machine at the end of period 3, so that we would still have a loss of $1800 at the end of period 3.

There are several figures of merit which could be used based on the same interest (discount) rate i, all of which would result in the same decision concerning whether or not to replace the machine. Since we must make our decision now, before the overhaul possibility, we will compute the discounted present worth (PW) at discount rate i, i.e., $PW(i)$. This can be calculated by bringing back to $t = 0$ the discounted cost of future actions. The present worth calculation if we do not replace is

$$PW(i) = -\$1500 - \$3000(1+i)^{-1} - \$3500(1+i)^{-2} - \$1800(1+i)^{-3}.$$

Cost of replacing with a new machine. Suppose the replacement cost of a new machine is $18,000 and maintenance costs for the new machine are $2500 per year for three years. The salvage value of the new machine at the end of three years will be only $6000 due to obsolescence.

Since we supposed the current salvage value of the old machine was $10,000 we have a total cost of only $8000 at time 0, while at the end of three years we have a salvage value minus operating costs for the third period of $6000 - \$2500 = \$3500 profit. Hence the present worth of the replacement decision based on our interest rate i is

$$PW(i) = -\$8000 - \$2500(1+i)^{-1} - \$2500(1+i)^{-2} + \$3500(1+i)^{-3}.$$

(See Figure 5.2.2.)

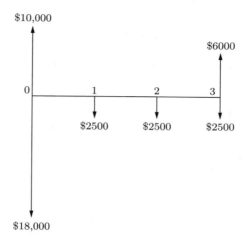

FIG. 5.2.2. *Cash flow if machine is replaced.*

TABLE 5.2.1
Keep or replace policies.

		Cost comparisons		
	Period	Keep old	Replace	Optimal policy
	0	−$1500	−$8000	
	1	−$3000	−$2500	
	2	−$3500	−$2500	
	3	−$1800	−$3500	
Total		−$9800	−$9500	Replace
PW(12%)		−$8249	−$9733	Keep old

To further bring our problem to life, suppose that our discount rate is $i = 0.12$ or 12% per year. This may represent our *minimum attractive rate of return*, i.e., the interest rate we could or at least would like to achieve. Using a spreadsheet such as EXCEL, for example, $PW(i)$ is easy to calculate in this and in much more complicated situations.

Table 5.2.1 summarizes our costs and calculations. The totals are the sums of costs without discounting.

Since the discounted cost at rate $i = 0.12$ (or interest rate 12%) of keeping the old machine for three years is $8249, while the discounted cost of replacing the old machine with a new machine is $9733.90, it is clear that our decision should be to *keep* the old machine.

Were we *not* to use discounting (i.e., $i = 0$) we would have computed $9800 for the cost of keeping the old machine and a cost of only $9500 for replacing the old with a new machine. Hence without discounting we would have decided to replace the old machine, whereas with discounting we would keep the old machine. Cost decisions over more than one year should be based on discounting.

Present worth versus future worth. It is conventional to calculate PW. However, costs are really not known with certainty but are *expected costs*. Hence we have calculated *expected PW* for each policy. This sounds wrong since it does not make sense to talk about what we expect to have today. We know what we have today. It is the future that is uncertain.

Expected future worth can be calculated using the factor $(1+i)^t$ instead of $(1+i)^{-t}$. Expected present worth and expected future worth will differ. However, as long as the discount rate i is specified and is the same we will make the same decision concerning policies using either criterion. If the discount rate is also uncertain, this is not true and we should use the future worth figure of merit.

Exercises

5.2.1. Using the data in Table 5.2.1 and an interest rate of $i = 0.10$, that is, 10%, determine the optimal policy regarding whether or not to keep the old machine.

5.2.2. Using the data in Table 5.2.1 and an interest rate of $i = 0.10$, compute the *annualized* expected cost per time period starting with period number 1. (By annualized we mean that the discounted cost is the same at the end of each period, usually one year, and has equivalent PW.) What is the optimal policy regarding whether or not to keep the old machine in this case?

5.2.3. Using the data in Table 5.2.1 and a random interest rate of either $i = 0.10$ or $i = 0.12$, each with equal probability, determine the optimal policy regarding whether or not to keep the old machine using the *expected future worth criterion*. What is the optimal decision now?

5.2.4. Why are optimal decisions relative to alternative decision comparisons based on future worth in general different from those based on present worth?

5.3. Sampling inspection.

In this section we discuss a sampling inspection problem which Deming (1986) made famous. Periodically, lots of size N of similar items arrive and are put into assemblies in a production line. The decision problem is whether or not to inspect items before they are put into assemblies. If we opt for inspection, what sample size n of the lot size N should be inspected? In any event, haphazard or skip-lot sampling to check on the proportion defective in particular lots is prudent.

Percent defective specified. Let π be the percent defective over many lots obtained from the same vendor. Suppose we believe that the vendor's production of items is in statistical control. A sequence of measurements will be said to be in statistical control if it is (infinitely) exchangeable. That is, each item, in our judgment, has the same chance π of being defective or good regardless of how many items we have already examined. Let $p(\pi)$ be our probability assessment for the parameter π based on previous experience. It could, for example, be degenerate at, say, π_o. Inspection results are independent only in the case that $p(\pi)$ is degenerate at π_o and π_o is specified.

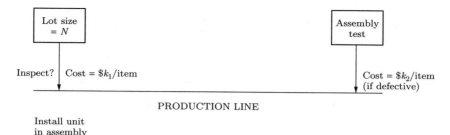

FIG. 5.3.1. *Deming's inspection problem.*

Let k_1 be the cost to inspect one item before installation. Let k_2 be the cost of a defective item that gets into the production line. This cost will include the cost to tear down the assembly at the point where the defect is discovered. If an item is inspected and found defective, additional items from another lot are inspected until a good item is found. (We make this model assumption since all items that are found to be defective will be replaced at vendor's expense.) Figure 5.3.1 illustrates our production line problem.

We assume the inspection process is perfect; i.e., if an item is inspected and found good then it will also be found good in the assembly test and if a defective item is not inspected, it will be found defective at assembly test.

The all or none rule. It has been suggested [cf. Deming, 1986, Chapter 15] that the optimal decision rule is always to either choose $n = N$ or $n = 0$, depending on the costs involved and π, the chance that an item is defective. Later we will show that this is not always valid when π is unknown.

In this example we consider the problem under the restriction that the initial inspection sample size is either $n = N$ or $n = 0$. Assuming that π *is known*, we cannot learn anything new about π by sampling. On the other hand, we can discover defective units by inspecting all N. Hence the only decisions to be considered are $n = N$ and $n = 0$. Figure 5.3.2 is an influence diagram for this problem, where

x = number of defectives in the lot and

y = number of additional inspections required to find good replacement items for bad items found either at the point of installation or the point of assembly test. These come from another lot.

The value (loss) function is

$$v[n, (x, y)] = \begin{cases} k_2 x + k_1 y & \text{if } n = 0, \\ k_1 N + k_1 y & \text{if } n = N. \end{cases}$$

If π is specified then inspection sampling cannot give us any new information about π. There are at least two reasons for inspection sampling:

(1) to find defective items,

(2) to learn about π when it is unknown.

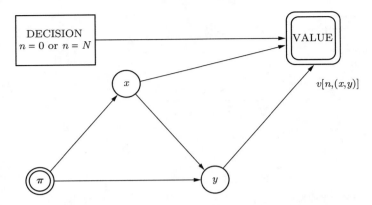

FIG. 5.3.2. *Influence diagram for Deming's inspection problem.*

If $n = N$ we will find all the defective items. If we perform sampling inspection to learn about π, then $n = 0$ since we are assuming that we already know π. Hence the optimal decision in this case is either $n = N$ or $n = 0$, depending on the costs involved. The double circle for π indicates that it is specified and hence deterministic. The foundations for influence diagrams such as the one in Figure 5.3.2 will be developed in Chapter 9. However, at this point it is not necessary to understand how to analyze such diagrams to solve our immediate problem.

To determine whether $n = 0$ or $n = N$ is best, we eliminate nodes x and y by calculating the expected value of the value function, first conditioning on π. The expected value given π is

$$E[v[n, (x, y)|\pi]] = \begin{cases} k_2 N\pi + k_1 E[y|\pi] & \text{if } n = 0, \\ k_1 N + k_1 E[y|\pi] & \text{if } n = N. \end{cases}$$

Hence $n = 0$ is best if

$$\pi < \frac{k_1}{k_2} \tag{5.3.1}$$

and $n = N$ is best if

$$\pi \geq \frac{k_1}{k_2}.$$

In the case that π is unknown and the only two decisions to be considered are either $n = N$ or $n = 0$, then the solution is also valid if we replace π by $E[\pi]$. If π or $E[\pi] = \frac{k_1}{k_2}$, then we may as well let $n = N$ since we may obtain additional information without additional expected cost.

Suppose now that we allow $0 \leq n \leq N$. If we are certain that $\pi < \frac{k_1}{k_2}$, i.e., $p(\pi) = 0$ for $\pi > \frac{k_1}{k_2}$, then $n = 0$ is always best. On the other hand, if we are certain that $\pi \geq \frac{k_1}{k_2}$ then $n = N$ is always best.

In the intermediate case, when $p(\pi)$ straddles

$$\pi = \frac{k_1}{k_2},$$

the optimal sample inspection size may be neither 0 nor N. In section 10.4, we consider the optimal solution when $0 \leq n \leq N$ can also be considered.

Value added to substrate. Work is done on incoming material, the substrate. The finished product will be classified as first grade, second grade, third grade, or scrap. If the incoming item is not defective, then it will result in a final product of first grade. Otherwise, if it is not inspected at the point of installation and if it is defective, then a final product of downgrade or scrap will be produced. Every final product is subject to assembly grading. The only difference between this model and the previous model is that no additional inspections occur at assembly test. This is because the item (say a bag of cement) cannot be replaced at assembly test if it is not inspected at the point of installation but is found defective at the point of assembly test. As before, let k_1 be the cost of inspecting an item. Let k_2 be the average loss from downgrading the final product or for scrapping finished items.

Again suppose that π, the percent defective over many lots, is specified. Then the only inspection sampling sizes we need to consider are $n = N$ or $n = 0$. Consider a typical item from the lot. If we inspect the item before installation we incur a cost k_1. If it is found to be defective, we must inspect additional items (but not from the same lot since the vendor must make good on any defective items). The expected number of additional items that must be inspected is $\frac{1}{1-\pi}$. If we inspect an item at the point of installation, the expected cost of inspection is

$$k_1 + k_1 \frac{\pi}{1 - \pi}.$$

If we do not inspect the item, the expected cost is $k_2\pi$. Hence $n = 0$ if

$$k_2\pi < k_1 + k_1 \frac{\pi}{1 - \pi}$$

or $n = 0$ if

$$\pi(1 - \pi) < \frac{k_1}{k_2}$$

and $n = N$ otherwise.

If

$$Z = \left\{ \begin{array}{ll} 1 & \text{if an item is defective,} \\ 0 & \text{otherwise,} \end{array} \right.$$

then $n = 0$ if

$$Var(Z|\pi) < \frac{k_1}{k_2} \qquad (5.3.2)$$

and $n = N$ otherwise.

The intuitive reason for the difference between (5.3.1) and (5.3.2) is as follows. If the variance is small, then π is either close to zero or close to one. If the variance is close to zero, there will be few defective items, so we should not inspect before installation. On the other hand, if we believe π is close to one, then there will be so many defective items that the cost of inspecting items before installation will tend to exceed the cost due to downgrading the item at the point of assembly test.

Exercises

5.3.1. Deming's inspection problem for a unique lot. Suppose we are considering whether or not to inspect items in a unique lot of size N. That is, there are no other items except those in this lot. Figure 5.3.1 still applies with some exceptions. There are *no additional inspections* in this case because the lot is unique.

Let θ be the (random) number defective in the lot of size N. If an inspection sample size of n is chosen and x are the number found defective in the sample, then we have a second decision to make, namely, either stop inspection or continue on and inspect all items.

(i) If the decision is to stop inspecting, the total inspection cost is $k_1 n + k_2(\theta - x)$.

(ii) If the decision is to continue on and inspect all remaining items, the total inspection cost is $k_1 N$.

If the prior distribution for θ is beta-binomial(N, A, B), determine the optimal decision after inspecting $n \leq N$ (specified) items; i.e., when is it optimal to stop inspection and when is it optimal to continue on when x defectives are found in a sample of size n?

Summary of discrete probability distributions. Let π be the chance that an item is defective.

Bernoulli

$$Z_i = \begin{cases} 1 & \text{with probability } \pi, \\ 0 & \text{with probability } 1 - \pi. \end{cases}$$

Binomial. If $Z_1, Z_2, \ldots, Z_n$ are judged independent given π, then

$$P(Z_1 + \cdots + Z_n = k | \pi) = \binom{n}{k} \pi^k (1 - \pi)^{n-k}$$

for $k = 0, 1, \ldots, n$ and $Z_1 + \cdots + Z_n$ is the number of bad items in a lot of size n. $Z_1 + \cdots + Z_n$ is binomial(n, π).

Geometric. Let W be the number of additional inspections required to find a good item after a bad item has been found. Then W has a geometric distribution

$$P(W = k | \pi) = \pi^{k-1}(1 - \pi) \qquad \text{for } k = 1, 2, \ldots$$

and

$$E[W|\pi] = \frac{1}{1 - \pi}.$$

Negative binomial. If $W_1, W_2, \ldots, W_x$ are judged independent given π and x, then $W(x) = W_1 + W_2 + \cdots + W_x$ will be the number of additional inspections required to find x good items:

$$P(W(x) = k|x, \pi) = \left(\begin{array}{c} n + k - 1 \\ k \end{array} \right) \pi^k (1 - \pi)^x.$$

The expected number of additional inspections required is

$$E[W(x)|x, \pi] = \frac{x}{1 - \pi}.$$

Beta-binomial with parameters (A, B, N) is the following:

$$p(x|A, B, N) = \frac{N! \Gamma(A + x) \Gamma(B + N - x) \Gamma(A + B)}{(N - x)! x! \Gamma(A + B + N) \Gamma(A) \Gamma(B)}$$

for $x = 0, 1, \ldots, N$. The mean is $E(X|N, A, B) = N \frac{A}{A+B}$.

5.4. Age replacement with discounting.

In this section we assume that items are from a conceptually infinite population with life distribution F. Under an age replacement policy we replace an item at failure or at the end of a specified time interval, whichever occurs first. This makes sense if a failure replacement costs more than a planned replacement and the failure rate is strictly increasing. We assume that the cost of a planned (failure) replacement is $c_1 (c_2)$ where $0 < c_1 < c_2$. These costs are assumed to be the same for each stage. (A stage is the period starting just after a replacement and ending just after the next replacement.)

Continuous discounting is used, with the loss incurred at the time of replacement and the total loss equal to the sum of the discounted losses incurred on the individual stages. Suppose that a stage starts at time t and we set a replacement interval T, where throughout we assume $T \in [0, \infty]$. If replacement actually occurs at $t + x$, then the loss incurred at that stage is

$$L(T, x, t) = \begin{cases} c_1 e^{-\alpha(t+T)} & \text{if } x = T, \text{ i.e., preventive replacement,} \\ c_2 e^{-\alpha(t+x)} & \text{if } x < T, \text{ i.e., failure replacement,} \end{cases}$$

where α is a positive discount (interest) rate.

We will assume in this section that the planning horizon is unbounded. The total risk at each stage is the same, except for the discount factor. For convenience, we compute the risk starting at time 0. Assuming that the failure distribution is specified and continuous, there exists a replacement interval that minimizes the risk. Since the optimal replacement age interval T does not depend on the starting time, there is some fixed replacement interval that is optimal to set at each stage. That is to say, there is a stationary and nonrandomized optimal policy.

Let F be the specified failure distribution with failure rate

$$r(x) = \frac{\left[\frac{dF(x)}{dx}\right]}{1 - F(x)}.$$

Let $\bar{F}(x) = 1 - F(x)$ and define

$$\phi(T) = c_1 e^{-\alpha T} \bar{F}(T) + c_2 \int_0^T e^{-\alpha x} dF(x),$$

$$\delta(T) = e^{-\alpha T} \bar{F}(T) + \int_0^T e^{-\alpha x} dF(x),$$

$$R(T) = \frac{\phi(T)}{[1 - \delta(T)]}, \qquad (5.4.1)$$

$$R(T^*) = \min_T R(T).$$

$R(T)$ is the long run expected discounted cost when the replacement interval T is set at each stage. If $r(x)$ is continuous and strictly increasing to ∞, then $T^* < \infty$ and is unique. Note that we need only specify the ratio of costs $\frac{c_1}{c_2}$ and not the costs separately.

If we take the limit

$$\lim_{\alpha \downarrow 0} \alpha R(\alpha, T) = \frac{c_1 \bar{F}(T) + c_2 F(T)}{\int_0^T \bar{F}(x)\, dx}, \qquad (5.4.2)$$

we obtain the long run average cost per unit of time. There is no discounting of costs, and the optimal replacement interval T minimizes long run average cost. If the failure rate $r(x)$ is strictly increasing to ∞, the minimizing $x = T^*$ satisfies

$$r(x) \int_0^x \bar{F}(u) du - F(x) = \frac{c_1}{c_1 + c_2}.$$

Note that we need only specify the ratio of costs $\frac{c_1}{c_2}$ and not the costs separately.

Example 5.4.1. To illustrate the optimal solution when $\lim_{\alpha \downarrow 0} \alpha R(\alpha, T)$, suppose $c_2 = 2c_1$ and $\bar{F}(x) = e^{-x^2}$, the Weibull distribution with shape parameter 2 and scale 1. The mean of the life distribution is 0.89 years or 10.71 months. We can use the connection with the cumulative normal distribution. Let

$$\Phi(x|\mu, \sigma^2) = \frac{1}{\sqrt{2\pi}\sigma} e^{-\frac{1}{2}\left(\frac{x-\mu}{\sigma}\right)^2}.$$

Then

$$\int_0^x e^{-u^2} du = \sqrt{\pi}\left[\Phi\left(x|\mu = 0, \sigma^2 = \frac{1}{2}\right) - \frac{1}{2}\right].$$

TABLE 5.4.1

Calculation of optimal replacement interval.

x in years	Cumulative normal	Left-hand side
0.1	0.54	0.03
0.2	0.58	0.04
0.3	0.62	0.09
0.4	0.66	0.17
0.5	0.69	0.28
0.53	0.70	0.32
0.535	**0.70**	**0.33**
0.54	0.71	0.34
0.55	0.71	0.35
0.6	0.73	0.43
0.7	0.76	0.63
0.8	0.79	0.90

Since $r(x) = 2x$, we have to solve the equation

$$2x\sqrt{\pi}\left[\Phi(x) - \frac{1}{2}\right] - \left[1 - e^{-x^2}\right] = \frac{1}{3}$$

to find T^*.

This can be done using a spreadsheet such as EXCEL, for example. Table 5.4.1 shows the calculations. From the table, $T^* = 0.535$ years or 6.42 months compared with the mean life of 10.71 months.

Exercises

5.4.1. Solve for T^* using

$$r(x)\int_0^x \bar{F}(u)du - F(x) = \frac{c_1}{c_1 + c_2}$$

with $c_2 = 2c_1$ and $\bar{F}(x) = e^{-x^3}$.

5.4.2. Using (5.4.1) prove (5.4.2).

5.5.   Notes and references.

Section 5.2. The material in this section is standard engineering economics. A good reference is Park (1997).

Section 5.3. Deming (1986) discusses the sampling inspection problem in Chapter 15 of his book. We follow the discussion in Barlow and Zhang (1986) and (1987).

Section 5.4. We follow the discussion of age replacement with discounting in Fox (1966).

PART **III**

System Reliability

In Chapters 6, 7, and 8 we assume failure and repair events are independent, conditional on component failure and repair distributions.

The purpose of a system reliability analysis is to acquire *information* about a system of interest relative to making decisions based on *availability, reliability,* and *safety*. In general it is best to be failure oriented in a system reliability analysis. There are several stages to such an analysis.

(1) *Inductive analysis stage.* Gather and organize available information. Ask the question, *What* can happen? Hypothesize and guess. This is the hard part of a system reliability analysis. At this stage we may perform a failure modes and effects analysis (FMEA) on critical components.

(2) *Deductive analysis stage.* This is the easier part. At this stage we ask the question, *How* can the system fail? To determine system reliability we use block (network) diagrams and fault tree construction.

General rules for failure prediction

(1) Look for human error possibilities.

(2) Failure often occurs at interfaces, i.e., where different elements and materials are joined.

A network and logic tree (or fault tree) representation is needed for reliability prediction and the calculation of availability. Although this is the easier part, we begin our discussion with a network model already formulated in the abstract and consider methods for calculating network reliability.

A block diagram is a system representation based on subsystems and/or critical components. In this representation, nodes may fail while arcs joining pairs of nodes are usually considered perfect. In our network reliability analysis, arcs may fail but nodes are usually considered perfect. A block diagram can also be represented as a network and vice versa.

CHAPTER 6

Network Reliability

6.1. Network reliability analysis: The factoring algorithm.

In this section and the next, system reliability is with respect to a specified instant in time. In Chapter 8 we extend our discussion of system reliability to time until system failure. The factoring algorithm of this section has the advantage that it begins directly with the network representation.

Consider the two terminal network in Figure 6.1.1. The *distinguished nodes* are the source node and the terminal node. The graph with nodes and arcs together with the distinguished nodes specifies the problem of interest. In this case, success corresponds to the event that there is at least one working path of nodes and arcs from source to terminal. To begin, we will assume nodes are perfect but that arcs can fail. Let

$$x_i = \begin{cases} 1 & \text{if arc } i \text{ works,} \\ 0 & \text{otherwise.} \end{cases}$$

Let the state vector corresponding to the state of arcs be

$$(x_1, x_2, \ldots, x_8).$$

Let

$$\phi(x_1, x_2, \ldots, x_8) = \begin{cases} 1 & \text{if the system works,} \\ 0 & \text{otherwise.} \end{cases} \tag{6.1.1}$$

Such abstract systems are among a class of systems called *coherent*. The *structure function* ϕ is the system organizing function that relates component operation to system operation.

DEFINITION. *A system of components is called* coherent *if* (a) *its structure function ϕ is increasing coordinatewise and* (b) *each component is relevant.*

A physical system in general (there are exceptions) has the property that replacing a failed component by a functioning component causes the system to pass from a failed state to a nonfailed state, or at least the system is no worse than before the replacement. This is our justification for requirement (a).

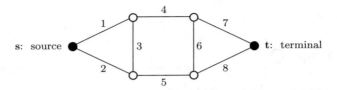

FIG. 6.1.1. *Two terminal network.*

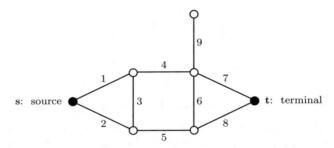

FIG. 6.1.2. *Example of an irrelevant arc.*

Component i is *irrelevant* if

$$\phi(1_i, x) = \phi(0_i, x)$$

for all choices of $(\bullet_i, x)$; i.e., it does not matter whether component i is working or not working. Component i is *relevant* otherwise. It follows that $\phi(1, 1, \ldots, 1) = 1$ since all components in a coherent system are relevant.

Figure 6.1.2 shows an example of a modified network that is not coherent because arc 9 is *not relevant* to the question regarding the existence of a working path from the source to the terminal. It may, however, be relevant in some other configuration of distinguished nodes.

Let $X_1, X_2, \ldots, X_8$ be binary random variables corresponding to arc success; i.e.,

$$X_i = \begin{cases} 1 & \text{if arc } i \text{ works,} \\ 0 & \text{otherwise.} \end{cases}$$

Given $P(X_i = 1) = p_i$ for $i = 1, \ldots, 8$ and assuming conditional independence given the p_i's, our problem is to compute the probability that there is at least one working path at a given instant in time between the distinguished nodes s and t. For a network with n arcs we will denote this probability by

$$h_\phi(\mathbf{p}) \stackrel{DEF}{=} h_\phi(p_1, p_2, \ldots, p_n). \tag{6.1.2}$$

It follows that $h_\phi(1, 1, \ldots, 1) = 1$ since all components in a coherent system are relevant.

The factoring algorithm for undirected networks. The factoring algorithm is the most efficient algorithm based on *series-parallel probability reduc-*

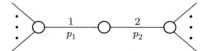

FIG. 6.1.3. *Simple series subnetwork.*

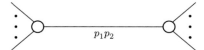

FIG. 6.1.4. *Reduced subnetwork.*

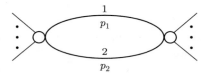

FIG. 6.1.5. *Simple parallel subnetwork.*

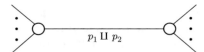

FIG. 6.1.6. *Reduced subnetwork.*

tions and *pivoting* for computing undirected network reliability when

(1) arcs fail independently of one another and

(2) nodes are deemed perfect.

The first idea is that of series-parallel probability reductions. Consider the simple series subnetwork in Figure 6.1.3. In this case, arcs 1 and 2 constitute a simple series subnetwork. Since arcs fail independently we can replace arcs 1 and 2 by a single arc with probability $p_1 p_2$ as in Figure 6.1.4.

Figure 6.1.5 is a simple parallel subnetwork. Since arcs fail independently, we can replace arcs 1 and 2 by a single arc with probability $p_1 \amalg p_2 = p_1 + p_2 - p_1 p_2$ as in Figure 6.1.6.

The symbol $\amalg$ is the "ip" operator where $p_1 \amalg p_2 = p_1 + p_2 - p_1 p_2$. In a computer program for calculating network reliability, the series and parallel probability reductions take relatively little computing time. On the other hand, the following pivoting operation does take appreciable computing time.

The probability basis for the pivoting operation follows from the identity $h(\mathbf{p}) = p_i h(1_i, \mathbf{p}) + (1 - p_i) h(0_i, \mathbf{p})$, valid when arcs fail independently. We use the notation $\mathbf{p} = (p_1, p_2, \ldots, p_n)$ and $(1_i, \mathbf{p}) = (p_1, p_2, \ldots, p_{i-1}, 1_i, p_{i+1}, \ldots, p_n)$.

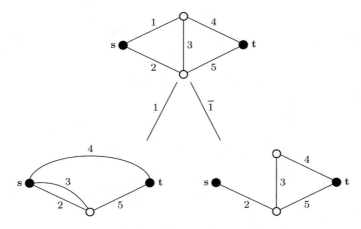

FIG. 6.1.7. *Illustration of the pivoting algorithm: the binary computational tree.*

To illustrate the pivoting operation, consider the "bridge" example in Figure 6.1.7. The top bridge network is the initial "node" in the illustrated algorithm. First note that the bridge network cannot be reduced to a single edge by series and parallel probability reductions. We pivot on arc 1. (Actually, we could just as well have pivoted on any other arc.) On the lower left we see the network with arc 1 perfect. On the lower right we see the network with arc 1 omitted. This diagram is the so-called binary computational tree illustrating our pivoting algorithm. The two networks at the bottom of the figure are the "leaves" of the binary computational tree. Were we to turn the diagram upside down, the top network "node" would become the "root" of the "tree" and the two bottom networks would look more like leaves. The two subnetworks created by the pivoting operation are called *minors* of the graph on which we pivot. In our illustration the leaves are the minors of the original network.

Since we can reduce both bottom networks by appropriate series and parallel probability reductions, we need not pivot again. Finally, we obtain the network reliability for the bridge

$$h(\mathbf{p}) = h(p_1, p_2, \ldots, p_5) = p_1 \left[((p_2 \amalg p_3)p_5) \amalg p_4 \right] + (1 - p_1)[p_2(p_3 p_4 \amalg p_5)].$$

If we let $p_1 = p_2 = \cdots = p_5 = p$ we have the *reliability polynomial*

$$h(p) = 2p^5 - 5p^4 + 2p^3 + 2p^2. \tag{6.1.3}$$

DEFINITION. *The reliability polynomial $h(p)$ is the network reliability function when we let all arc success probabilities equal p.*

The reliability polynomial for coherent systems can also be used to obtain bounds on system reliability when component reliabilities differ. Let

$$p_{\min} = \min\{p_1, p_2, \ldots, p_n\}$$

and

$$p_{\max} = \max\{p_1, p_2, \ldots, p_n\}.$$

Then

$$h(p_{\min}) \leq h(\mathbf{p}) \leq h(p_{\max}). \tag{6.1.4}$$

This is because $h(\mathbf{p})$ is coordinatewise nondecreasing.

DEFINITION. *The coefficient of p^n in the reliability polynomial is called the* signed domination. *The absolute value of the coefficient of p^n is called the* domination.

The absolute value of the coefficient of p^n in the reliability polynomial for an undirected network with n arcs corresponds to the number of leaves in the binary computational tree for the factoring algorithm.

In the case of our bridge example, $n = 5$ and the coefficient of p^5 in (6.1.3) is 2. There are exactly two leaves in the binary computational tree in Figure 6.1.7. We stop with two leaves in this case since each can be reduced to a single edge by series-parallel probability reductions.

It can be shown that the number of "nodes" in the binary computational tree is $2L - 1$, where L is the number of leaves. Consequently, the number of pivots required is $(2L - 1) - L = L - 1$.

The factoring algorithm. To implement the factoring algorithm, first delete all irrelevant arcs. Then perform the following steps:

(1) Perform all possible series-parallel probability reductions.

(2) Pivot on an edge such that the minors created are coherent (i.e., contain no irrelevant arcs).

(3) Again perform all possible series-parallel probability reductions on the minors created.

(4) Continue in this fashion until the only minors left are series-parallel graphs. These are the leaves of the binary computational tree.

(5) Calculate network reliability by calculating the reliability of each of the leaves weighted by the sequence of probabilities leading to their creation.

Exercises

6.1.1. What is the sum of the coefficients of the reliability polynomial for an arbitrary undirected network with n arcs?

6.1.2. Calculate the *reliability polynomial* for the two terminal undirected network in Figure 6.1.1 using the factoring algorithm. Show your work.

6.1.3. Use the factoring algorithm to calculate the reliability polynomial for the *all terminal* undirected network below, i.e., the probability that any node can communicate with any other node in the network when all arc success probabilities are set equal to p. In this case all nodes are distinguished and each needs to communicate with all other nodes.

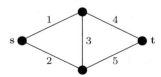

6.1.4. The network reliability problem for the figure below is source to terminal connectivity and arcs are unreliable.

(a) Find the domination and signed domination for the undirected network below.

(b) Is this network coherent?

(c) Using the factoring algorithm, would it be permissible to first pivot on arc 6 and then on arc 7? Why or why not?

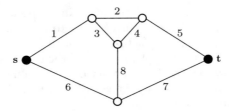

6.1.5. Show that the domination of the undirected two-terminal graph below is 2^{n-1}. The point of this example is to show that the factoring algorithm is *not* a polynomial time algorithm. The intermediate nodes between s and t are labelled.

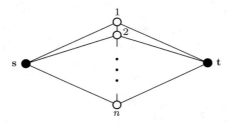

6.1.6. Using the definition of coherent systems, prove that $h(1, 1, \ldots, 1) = 1$.

6.2. General methods for system reliability calculation.

In the previous section we assumed that arcs failed independently and that nodes were perfect. This need not be so, and therefore more general methods are required.

Boolean reduction. A very general method called Boolean reduction starts with the minimal path and/or minimal cut sets for the network. In general, finding the minimal path or minimal cut sets is NP-hard in the language of computer complexity theory. This means that for a network with n arcs, the computing time necessary to find these sets is, in the worst case, exponential in n rather than, say, polynomial time in n.

Consider the "bridge" example in Figure 6.2.1. For the bridge, the *minimal path sets* are

$$P_1 = \{1, 4\}, \quad P_2 = \{2, 5\}, \quad P_3 = \{1, 3, 5\}, \quad P_4 = \{2, 3, 4\}. \quad (6.2.1)$$

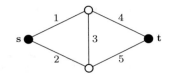

FIG. 6.2.1. *The "bridge."*

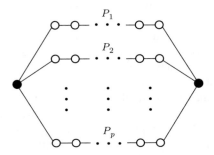

FIG. 6.2.2. *Minimal path representation.*

For example, if arcs 1 and 4 in P_1 are working but all other arcs are failed, then there is still a working path from source **s** to terminal **t**. In addition, the set P_1 cannot be reduced and still constitute a working path. So $\{1, 4, 3\}$ would also constitute a path set but would not be minimal.

Likewise for the bridge, the *minimal cut sets* are

$$K_1 = \{1, 2\}, \quad K_2 = \{4, 5\}, \quad K_3 = \{1, 3, 5\}, \quad K_4 = \{2, 3, 4\}. \qquad (6.2.2)$$

If arcs 1 and 2 are failed, for example, then there is no path from **s** to **t** regardless of the state of other arcs and K_1 is called a cut set. As in the case of path sets, we are only interested in minimal cut sets.

Using either the minimal path sets or the minimal cut sets we can define a Boolean representation for a network that, although artificial, provides another method for calculating network reliability.

The minimal path representation. Let $P_1, P_2, \ldots, P_p$ be the minimal path sets for a connected network with distinguished nodes. Then we can diagram a minimal path set representation as in Figure 6.2.2. It follows that

$$\phi(\mathbf{x}) = \phi(x_1, x_2, \ldots, x_n) = \coprod_{j=1}^{p} \prod_{i \in P_j} x_i, \qquad (6.2.3)$$

where $\Pi_{i \in P_j} x_i$ is the system structure function for the jth minimal path set. There are p minimal path sets. Minimal path sets are in turn connected in parallel. The representation is not necessarily physically realizable since the same arc (or component) can occur in several minimal path sets.

A similar representation holds with respect to minimal cut sets where now the network system structure function is

$$\phi(\mathbf{x}) = \phi(x_1, x_2, \ldots, x_n) = \prod_{j=1}^{k} \coprod_{i \in K_j} x_i \qquad (6.2.4)$$

and k is the number of minimal cut sets $K_1, K_2, \ldots, K_k$.

To calculate the reliability function using Boolean reduction,

(1) expand the Boolean expression, possibly obtaining integer powers of Boolean indicators;

(2) replace powers of Boolean indicators by the indicator itself; e.g., replace x_i^m by x_i wherever a power of x_i occurs thus obtaining a multilinear expression;

(3) if arcs fail independently, replace x_i wherever it occurs in the multilinear expression by p_i to obtain the network reliability;

(4) if arcs do not fail independently, take the expectation of the multilinear expression after replacing powers of Boolean indicators by the corresponding random indicator variable. For example, $E(X_i X_j)$ may have to be calculated using the joint probability function for both X_i and X_j.

Although the factoring algorithm does not work in the following directed network reliability problem, the Boolean reduction method does.

The factoring algorithm does not work for rooted directed graphs. A directed graph (or digraph) has arcs with arrows indicating direction. Figure 6.2.3 is a rooted directed graph. The "root" is vertex **s**, the unique node with no entering arcs. The problem is to compute the probability that **s** can communicate with both **v** and **t**. The topology for this problem is defined by the family of minimal path sets

$$\mathbf{P} = [\{2,5\}, \{1,3,5\}, \{2,e,4\}, \{1,3,4\}, \{1,2,4\}]. \qquad (6.2.5)$$

Graphically, a minimal path for this problem is a "rooted tree." For example, the acyclic rooted directed graph in Figure 6.2.4 is a "tree" corresponding to a minimal path for the network in Figure 6.2.3. Unfortunately, the factoring algorithm is not valid for this problem or for directed graphs in general. The binary computational tree shown in Figure 6.2.5 illustrates what goes wrong.

Note that arc 3 is now a self-loop after nodes **u** and **v** are coalesced. If arc e is perfect, then the corresponding minimal path sets are

$$\{2,5\}, \quad \{1,3,5\}, \quad \{2,4\},$$

whereas from the leaf graph on the left we would infer that $\{1,4\}$ is also a minimal path which is FALSE!

Domination and the principle of inclusion–exclusion: Back to the factoring algorithm. Let $C = \{1, 2, \ldots, n\}$ be the components (or arcs in the case of a network) for an arbitrary coherent system. Let $[P_1, P_2, \ldots, P_p]$ be the

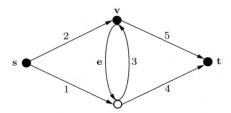

FIG. 6.2.3. *Directed network with three distinguished nodes.*

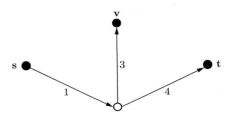

FIG. 6.2.4. *A rooted tree for Figure 6.2.3.*

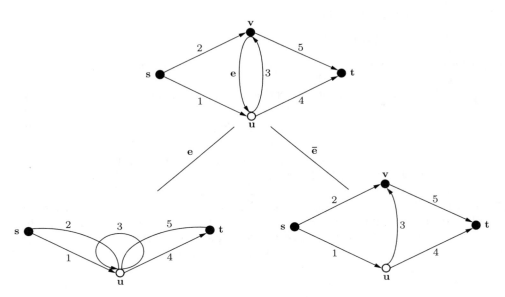

FIG. 6.2.5. *Pivoting on arc e.*

minimal path sets for the coherent system. Let E_r, $r = 1, 2, \ldots, p$ be the event
that all components (or arcs in the case of a network) in a minimal path set of
the coherent system work. By the inclusion–exclusion principle, we can calculate

system reliability

$$h(\mathbf{p}) = \sum_{r=1}^{p} P(E_r) - \sum_{i<j} P\left(E_i \bigcap E_j\right) + \cdots + (-1)^{p-1} P\left(E_1 \bigcap E_2 \bigcap \cdots \bigcap E_p\right),$$
$$(6.2.6)$$

where $E_i \cap E_j$ is the event that both E_i and E_j occur.

A *formation* is a set of minimal path sets whose union is the set of all components (or arcs in the case of networks). For example, in the case of the bridge with minimal path sets given by (6.2.1)

$$P_1 = \{1,4\}, \quad P_2 = \{2,5\}, \quad P_3 = \{1,3,5\}, \quad P_4 = \{2,3,4\}, \qquad (6.2.1)$$

all the formations of $C = \{1,2,3,4,5\}$ are

$F_0 = [\{1,4\},\{2,5\},\{1,3,5\},\{2,3,4\}],$

$F_1 = [\{2,5\},\{1,3,5\},\{2,3,4\}],$

$F_2 = [\{1,4\},\{2,5\},\{2,3,4\}],$

$F_3 = [\{1,4\},\{2,5\},\{1,3,5\}],$

$F_4 = [\{1,3,5\},\{2,3,4\}],$

$F_5 = [\{1,4\},\{1,3,5\},\{2,3,4\}].$

A formation is said to be even (odd) if the number of minimal path sets is even (odd).

In the reliability polynomial for the bridge, we repeat equation (6.1.3):

$$h(p) = 2p^5 - 5p^4 + 2p^3 + 2p^2. \qquad (6.1.3)$$

The coefficient of p^5, where $n = 5$ in this case, corresponds to the number of odd formations of $C = \{1,2,3,4,5\}$ minus the number of even formations of $C = \{1,2,3,4,5\}$; i.e., $4 - 2 = 2$. We call this the signed domination, and the absolute value of this coefficient we call the domination. In general, the coefficient of p^n in the reliability polynomial corresponds to the signed domination and the absolute value of this coefficient is the domination. The domination is a measure of computational complexity for the factoring algorithm.

Exercises

6.2.1. Use Boolean reduction to calculate the reliability for the network in Figure 6.2.3.

6.2.2. What is the domination number for the directed graph in Figure 6.2.3?

6.2.3. Show that $h(\mathbf{p})$ is nondecreasing coordinatewise for coherent systems.

6.3. Analysis of systems with two modes of failure: Switching and monitoring systems.

Switching systems can not only fail to close when commanded to close but can also fail to open when commanded to open. Likewise, safety monitoring systems can not only fail to give warning when they should but can also give false alarms. In either case, the switching or the monitoring systems are not working properly. It is desirable to design switching and monitoring systems that are reliable and also do not have too many failures of the second type. The design of such systems can be analyzed using coherent system duality and the reliability polynomial.

Dual coherent systems. Duality in mathematics is a principle whereby one true statement can be obtained from another by merely interchanging two words. In the case of a coherent system, the dual coherent system can be obtained from the coherent system by interchanging the OR and AND operations. When we do this, minimal path sets for the primal system, say, ϕ, become minimal cut sets for the dual system called ϕ^D. To see this, recall the minimal path and minimal cut set representations for coherent systems:

$$\phi(\mathbf{x}) = \phi(x_1, x_2, \ldots, x_n) = \coprod_{j=1}^{p} \prod_{i \in P_j} x_i = \prod_{j=1}^{k} \coprod_{i \in K_j} x_i,$$

where P_j and K_j are minimal path and minimal cut sets, respectively. If we interchange the OR (i.e., $\amalg$) and AND (i.e., Π) operations to get ϕ^D, we have

$$\phi^D(\mathbf{x}) = \prod_{j=1}^{p} \coprod_{i \in P_j} x_i = \coprod_{j=1}^{k} \prod_{i \in K_j} x_i, \tag{6.3.1}$$

where now P_j is a minimal cut set for ϕ^D and K_j is a minimal path set for ϕ^D. It follows that

$$\phi^D(\mathbf{x}) = 1 - \phi(\mathbf{1} - \mathbf{x}) \tag{6.3.2}$$

since

$$\phi^D(\mathbf{x}) = \prod_{j=1}^{p} \coprod_{i \in P_j} x_i = 1 - \coprod_{j=1}^{p} \prod_{i \in P_j} (1 - x_i) = 1 - \phi(\mathbf{1} - \mathbf{x}). \tag{6.3.3}$$

Example 6.3.1. 2-out-of-3 systems. Suppose a warning system consists of three exchangeable smoke detectors placed in close proximity but operating independently. Suppose furthermore that we construct our alarm system in such a way that an alarm sounds at a listening post if and only if at least two of the detectors are triggered. The minimal path sets are

$$P_1 = \{1, 2, \}, \quad P_2 = \{1, 3\}, \quad P_3 = \{2, 3\},$$

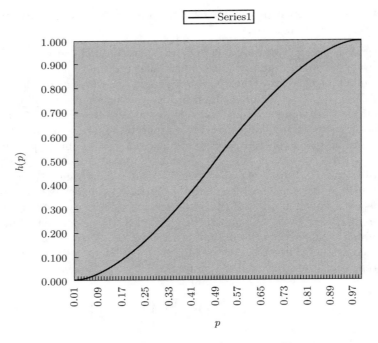

FIG. 6.3.1.   *Reliability polynomial for a 2-out-of-3 system.*

and the minimal cut sets are also

$$K_1 = \{1,2\}, \quad K_2 = \{1,3\}, \quad K_3 = \{2,3\},$$

so that the coherent structure function in this case is self-dual; i.e.,

$$\phi^D(\mathbf{x}) = \phi(\mathbf{x}).$$

Let p be the probability that a detector alarms when smoke is present. Then the reliability polynomial for the 2-out-of-3 alarm system is

$$h(p) = \sum_{i=2}^{3} \binom{3}{i} p^i (1-p)^{3-i} = 3p^2(1-p) + p^3.$$

Figure 6.3.1 is a graph of the reliability polynomial. Since ϕ in this case is self-dual we have

$$h_{\phi^D}(p) = h_\phi(p),$$

which in turn implies $1 - h_\phi(1-p) = h_\phi(p)$ from (6.3.2). Note that $h(p)$ is S-shaped about $p = 0.5$ in this case.

The dual coherent system and two modes of failure.   First we consider a switching system which is subject to two modes of failure: failure to close and failure to open. Similarly, circuits constructed of switches (to mitigate such failures) are subject to the same two modes of failure.

Let ϕ be a coherent switching system with n exchangeable and statistically independent components. Let $x_i = 1$ if the ith switch correctly closes when commanded to close and $x_i = 0$ otherwise. Let $\phi = 1$ if the circuit responds correctly to a command to close and 0 otherwise. Then $\phi(\mathbf{x})$ represents the structure function representing the correct system response to a command to close.

Now let $y_i = 1$ if the ith switch responds correctly to a command to open (that is, opens) and 0 otherwise. Let ψ be a similar structure function such that $\psi = 1$ if the circuit responds correctly to a command to open (that is, opens) and 0 otherwise. Then

$$\psi(\mathbf{y}) = \phi^D(\mathbf{y}) = \prod_{j=1}^{k} \prod_{i \in K_j} y_i \qquad (6.3.4)$$

from (6.3.1) and since the minimal cuts $\{K_j | j = 1, 2, \ldots, k\}$ for ϕ are the minimal paths for ψ. Recall that a minimal cut set of switches for ϕ is a set of switches whose failure to close means the switch system cannot close and therefore the system is open when the switches in a minimal cut set for ϕ are open.

Let $X_i = 1$ with probability p so that

$$P[\phi(\mathbf{X}) = 1 | p] = h_\phi(p).$$

Let $Y_i = 1$ with probability p' so that

$$P[\psi(\mathbf{Y}) = 1 | p'] = h_{\phi^D}(p') = 1 - h_\phi(1 - p').$$

Consequently, when $p' \approx 1$ we want $h_\phi(1 - p') \approx 0$ or $h(p) \approx 0$ when $p \approx 0$. Hence the S-shaped characteristic of 2-out-of-3 systems, for example, is exactly what we need.

By $h_\phi(p)$ being S-shaped we mean that there is a value $p_0 (0 < p_0 < 1)$ such that $h_\phi(p_0) = p_0$ and $h_\phi(p) < p$ for $0 < p < p_0$ and $h_\phi(p) > p$ for $p_0 < p < 1$. Although this is generally true for coherent systems, it is most pronounced for k-out-of-n systems.

Exactly the same reasoning applies to monitoring systems where monitoring detectors can fail to alarm when a threat such as smoke is present as well as false alarm when a threat is not present.

Computing k-out-of-n system reliability. Coherent systems that are k-out-of-n are especially useful in controlling the two modes of failure considered above. A system is k-out-of-n if it works whenever k or more components work. Although we have previously assumed that all components are exchangeable, this is not necessary. Let p_i, $i = 1, 2, \ldots, n$ be the probability that component i closes (alarms) when commanded to close (when a threat is present). To calculate

$h_\phi(\mathbf{p})$ use the generating function

$$g(z) = \prod_{i=1}^{n}(q_i + p_i z),$$

where $q_i = 1 - p_i$.

Expand and sum coefficients of z^j for $j = k, \ldots, n$ to obtain the probability that at least k components operate. Computing time is of the order $\frac{n^2}{2}$ for fixed k. A computer program written in BASIC follows:

```
100   REM K-OF-N: STRUCTURAL RELIABILITY
210   INPUT "K,N=";K,N: DIM A(N+1): A(1)=1
510   FOR J=1 TO N
520       PRINT "P(";J;")=";: INPUT P
551       FOR I=J+1 TO 1 STEP -1
555           A(I)=A(I)+ (A(I-1)-A(I))*P
559       NEXT I
590   NEXT J
810   FOR J=K+1 TO N+1: S=S+A(J): NEXT J
910   PRINT "STRUCTURE RELIABILITY IS:"S
990   END
```

After program execution, $A(J + 1)$ is the probability that exactly J components operate.

Exercises

6.3.1. Verify equation (6.3.3).

6.3.2. Calculate and graph $h_\phi(p)$ when ϕ is a 4-out-of-5 system. Use $h_\phi(p)$ to calculate and graph the dual system.

6.3.3. Show that the dual of a k-out-of-n system is an $n - k + 1$-out-of-n system.

6.3.4. Use the BASIC program above to calculate $h_\phi(p)$ when ϕ is a 3-out-of-5 system with component reliabilities

$$p_1 = 0.10, \quad p_2 = 0.30, \quad p_3 = 0.50, \quad p_4 = 0.70, \quad p_5 = 0.80.$$

6.3.5. A safety alarm system is to be built of four identical, independent components. The following structures are under consideration:

(i) 3-out-of-4 system; i.e., the system works if three or more components work;

(ii) 2-out-of-4 system;

(iii) "1-out-of-2-twice system; i.e., see the following diagram.

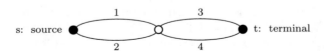

(a) Assuming each component has reliability p, compute the reliability function for each structure.

(b) Which system is most reliable for p close to 1?

(c) Which system is most prone to false alarms when the probability of a component false alarm is very low?

6.4. Notes and references.

Section 6.1. Coherent systems were first defined and discussed by Birnbaum, Esary, and Saunders (1961). The factoring algorithm is due to Chang and Satyanarayana (1983). The role of domination in the factoring algorithm is discussed in Barlow and Iyer (1988).

Section 6.2. See Colbourn (1987) and Rai and Agrawal (1990) for more recent additional material on network reliability.

Section 6.3. Moore and Shannon (1956) were the first to show the "sharpness" of the k-out-of-n systems. The program for computing k-out-of-n reliability appears in Barlow and Heidtmann (1984).

Notation

ϕ	system structure function
$\boldsymbol{p}$	$\boldsymbol{p} = (p_1, p_2, \ldots, p_n)$
$h_\phi(\boldsymbol{p})$	reliability function for system ϕ
$\amalg$	$p_1 \amalg p_2 = p_1 + p_2 - p_1 p_2$
$\amalg$	$\amalg_{i=1}^{2} p_i = p_1 + p_2 - p_1 p_2$
$\prod$	$\prod_{i=1}^{n} p_i = p_1 p_2 \cdots p_n$
$\amalg_{j=1}^{p} \prod_{i \in P_j} x_i$	minimal path representation of ϕ
$\prod_{j=1}^{k} \amalg_{i \in K_j} x_i$	minimal cut representation of ϕ
ϕ^D	dual of ϕ: $\phi^D(\mathbf{x}) = 1 - \phi(\mathbf{1} - \mathbf{x})$

CHAPTER 7

System Failure Analysis: Fault Trees

7.1. System failure analysis.

Network reliability is based on an abstract graphical representation of a system. It is basically success oriented. In practice, as it turns out, it is best to be failure oriented. Once this position is taken, the number of events that can be thought about that could cause or result in system failure becomes very large. Hence it is necessary to set spatial and temporal bounds on the system. It is necessary to determine only those events absolutely relevant to the analysis.

A fault tree (or logic tree) is often the best device for deducing how a major system failure event could possibly occur. Since the network graph is close to a system functional representation, it cannot capture abstract system failure and human-error events as well as the fault tree representation. The fault tree is a special case of a class of diagram representations called *influence diagrams*, developed in Chapter 9. In sections 7.2 and 7.3 we develop the background necessary for constructing and analyzing fault tree diagrams. However, their construction first depends on a thorough understanding of the system and the results of a system inductive analysis.

Simulation versus fault tree analysis. Currently, simulation is perhaps more popular than fault tree analysis. Faster and faster computers make simulation very attractive. It also has the advantage that the dynamics of system operation can be studied, whereas the fault tree is essentially a static system representation. However, like the network representation, simulation tends to be success oriented. Possible but rare fault events may be overlooked or inadequately treated in simulation. Hence we still recommend that the fault tree approach be used first, followed by simulation. Fault tree has the advantage of concentrating attention on learning in depth about the system of interest.

System inductive analysis. Initially we need to perform a system inductive analysis before we construct networks, fault trees, influence diagrams or employ simulation. This inductive analysis consists of
  (1) a system description,
  (2) a preliminary gross subsystems hazard analysis,

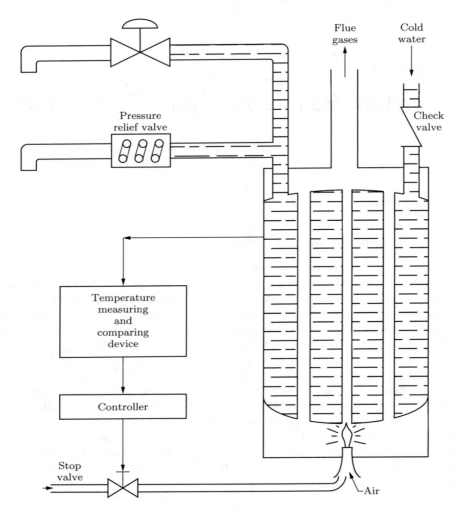

FIG. 7.1.1. *Schematic of a domestic hot water system.*

(3) a component failure modes and effects analysis (FMEA).

The system should be specified in terms of

(a) its functional purpose, time period of use, and environmental conditions to be encountered;

(b) its components as well as the people involved;

(c) the functional order of the system such as inputs, outputs, the logic of its operation, etc.

A very useful outline of the inductive analysis phase of a system analysis is given by the following safety analysis relative to possible rupture of a hot water tank; see Figure 7.1.1.

Outline of a system description: A domestic hot water system.

1. *Functional purpose of the system*
   (a) *What?* Provide a supply of water heated to a preset temperature for household use on an automatic basis.
   Specify
   - Volume stored
   - Inlet and outlet temperature limits
   - Inlet pressure limits
   - Recovery rate
   - Gas consumption (type of gas) limits
   (b) *When?* For a system lifetime of 20 years at 100% duty, etc.
   (c) *Where?* Domestic household environment.
   Specify
   - Temperature extremes
   - Location (indoor, outdoor, etc.)
   - Ventilation

2. *Component identification*
   (a) *Subsystems.* For example
       (i) Temperature measuring and controlling device
       (ii) Gas burner
       (iii) Storage and protective
       (iv) Distribution
   (b) *Components.* For example
       (i) Thermostatic device, controller, gas valve, piping, etc.
       (ii) Burner casting, air mixture control, pilot burner, etc.
       (iii) Water tank pressure relief valve, flue pipe, check valve
       (iv) Piping, faucets, etc.

3. *Functional order of the system*
   (a) The gas valve is operated by the controller which in turn is operated by a temperature measuring and comparing device. The gas valve operates the main burner in full-on/full-off modes.
   The check valve in the water inlet prevents a reverse flow due to over-pressure in the hot water system.
   The pressure relief valve opens when pressure in the system exceeds 100 psi.
   (b) When the temperature of the water is below a preset level (140°F to 180°F), the temperature measuring and comparing device signals the controller to open the gas valve and turn on the main gas burner which is lit by the pilot burner. When the water temperature reaches a preset level, the temperature measuring and comparing device signals the controller to turn off the gas valve and thus turns off the main gas burner.

Undesired system events. Before performing a deductive (fault tree) analysis, we need to understand what the undesired system events are from various standpoints, namely,

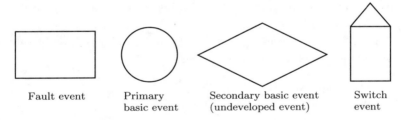

Fault event   Primary   Secondary basic event   Switch
                     basic event   (undeveloped event)   event

FIG. 7.2.1. *Event symbols.*

1. *Safety*
   (a) Fire due to a system fault
   (b) Explosion due to a system fault
       (i) Tank rupture
       (ii) Gas air explosion
   (c) Toxic effects
       (i) Gas asphyxiation
       (ii) Carbon monoxide poisoning
2. *Reliability*
   (a) Failure to supply heated water
   (b) Water supplied at excess temperature
   (c) Water supplied at excess pressure

7.2. Fault tree construction.

Fault tree construction and analysis is one of the principle methods of systems
safety analysis. It evolved in the aerospace industry in the early 1960s. Fault
tree analysis can be a valuable design tool. It can identify potential accidents in
a system design and can help eliminate costly design changes and retrofits. It
can also be a diagnostic tool that can predict the most likely causes of system
failure in the event of system breakdown.

A fault tree is a model that graphically and logically represents the various
combinations of possible events, both fault and normal, occurring in a system
that can lead to the "top event." The term "event" denotes a dynamic change
of state that occurs in a system element. System elements include hardware,
software, as well as human and environmental factors.

Event symbols.   The symbols shown in Figure 7.2.1 represent specific types
of fault and normal events in fault tree analysis. The rectangle defines an event
that is the output of a logic gate and is dependent on the type of logic gate and
the inputs to the logic gate. A "fault event" is an abnormal system state. It is
not necessarily due to a component failure event. For example, the fault event
could occur due to a command or a communication error.

The circle defines a basic inherent failure of a system element when operated
within its design specifications. We refer to this event as a primary basic event or
"end of life" failure event. The diamond represents a failure, other than a primary

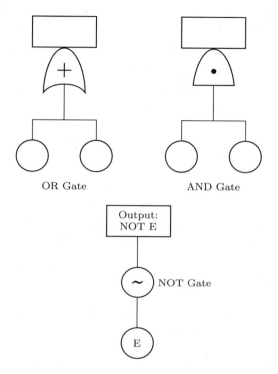

FIG. 7.2.2. *Logic gates.*

failure that is purposely not developed further. We call this a secondary basic
event. Basic events are either primary events (denoted by a circle) or secondary
basic events (denoted by a diamond). The switch event represents an event that
is expected to occur or to never occur because of design and normal conditions
such as a phase change in the system.

Logic gates. Fault trees use "OR gates" and "AND gates." OR gates are
inclusive in the sense that the output occurs if one or more input events occur.
The output event from an AND gate occurs only if all input events occur. In
addition, there may be a NOT gate in a fault tree. A NOT gate has a single
input event. The output occurs if the input event does not occur. Any logical
proposition can be represented using OR and AND together with NOT logic
symbols.

Additional logic gates. In addition to the principal logic gates in Figure 7.2.2, it is sometimes useful to use a "PRIORITY AND" gate as well as
a so-called EXCLUSIVE OR gate. For a PRIORITY AND gate, the output
event occurs if and only if the input events occur in time sequence from the left
to the right. Figure 7.2.2 (continued) has a PRIORITY AND gate as well as an
EXCLUSIVE OR gate.

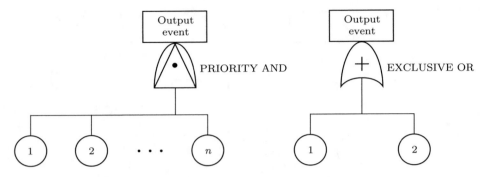

FIG. 7.2.2 (*continued*).   *Logic gates.*

The output event from an EXCLUSIVE OR gate occurs in the case of two input events if either one input occurs or the other input occurs but not both.

Steps in a fault tree analysis.   The following outline describes the steps to be taken in a fault tree analysis. The sequence of events leading to the top event is called a cut set.

    Step 1. Define the top event (usually an undesired system event).

    Step 2. Acquire an understanding of the system (this is the inductive analysis phase).

    Step 3. Construct the fault tree (this is the deductive analysis phase).

    Step 4. Evaluate the fault tree

       (a) Qualitative

          Visual inspection

          Find minimal cut sets

       (b) Quantitative

          Find the Probability of the top event

          Identify "important" components or basic events

          Rank basic events by importance

The fault tree in Figure 7.2.3 for the top event "rupture of a hot water tank" is an elaborate illustration of a fault tree together with notes explaining the events. The fault tree is developed using engineering knowledge and judgment. There may be many fault tree representations adequate for a given system analysis.

The following useful rules for fault tree construction were first formulated by D. F. Haasl and published in the "Fault Tree Handbook" by the U.S. Nuclear Regulatory Commission as NUREG-0492 in 1981 (see Vesely et al.).

Rules for fault tree construction.

    I. Starting with the top event, determine the *immediate necessary and sufficient causes* for the occurrence of the top event. It should be noted that these are not the basic causes of the event but the immediate causes or immediate mechanisms for the event.

    II. Write the statements that are entered in the event boxes as faults; state precisely what the fault is and when it occurs. *A fault event is not necessarily*

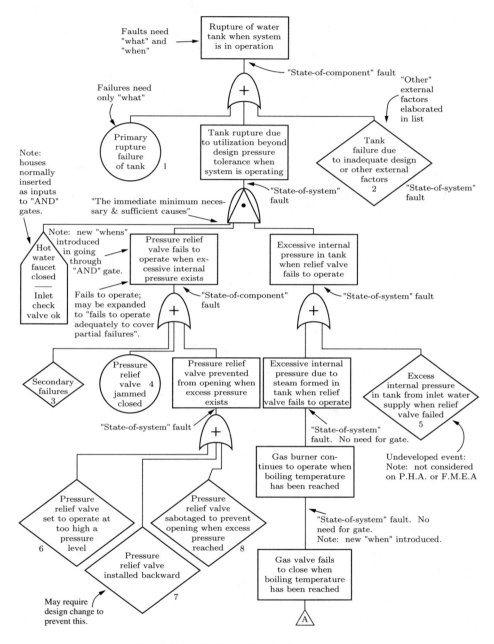

FIG. 7.2.3. *Fault tree for the hot water tank.*

a failure event. It may just be that a mechanism has been commanded to do something that it should not, in which case the mechanism is at fault but may not have failed.

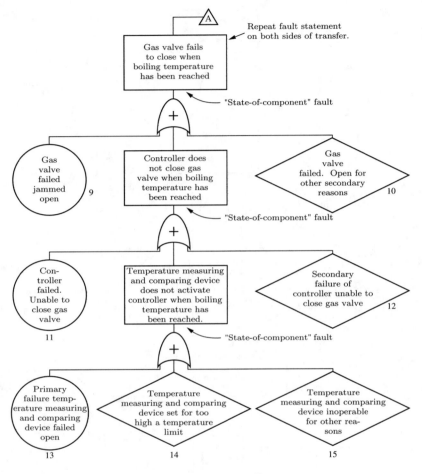

FIG. 7.2.3 (*continued*).

III. For each event in a rectangle we ask the question, Can this fault consist of a component failure? If the answer is *Yes*, classify the event as a "state-of-component fault." Add an OR gate below the event and look for primary, secondary, and command faults.

If the answer is *No*, classify the event as a "state-of-system" fault event. In this case, the appropriate gate could be OR, AND, or perhaps no gate at all. In the last case, another rectangle is added to define the event more precisely.

IV. *The no miracles allowed rule.* If the normal functioning of a component propagates a fault sequence, then it is assumed that the component functions normally.

V. All inputs to a particular gate should be completely defined before further analysis of any one of them is undertaken.

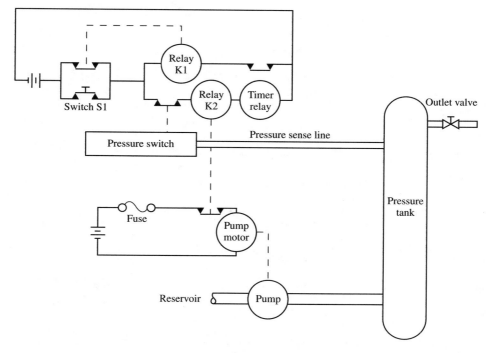

FIG. 7.2.4. *Pressure tank design.*

VI. *The no gate-to-gate connections rule.* Gate inputs should be properly defined fault events, and gates should not be directly connected to other gates.

The pressure tank fault tree example. Consider the pressure tank diagram in Figure 7.2.4. To upgrade system safety we want to develop a fault tree in which the top event is a pressure tank rupture.

In the operational mode, to start the system pumping, the reset switch S1 is closed and then opened immediately. This allows current to flow in the control branch circuit, activating relay coil K2. K2 contacts close and start the pump motor.

After approximately 20 seconds, the pressure switch contacts should open (since excess pressure should be detected by the pressure switch), deactivating the control circuit, de-energizing the K2 coil, opening the K2 contacts, and thereby shutting the motor off. If there is a pressure switch hang-up (emergency shutdown mode), the timer relay contacts should open after 60 seconds, de-energizing the K2 coil and shutting off the pump. We assume that the timer resets itself automatically after each trial, that the pump operates as specified, and that the tank is emptied of fluid after every run.

The fault tree is shown in Figure 7.2.5. The top event, a pressure tank rupture, is a state-of-component fault. The tree is developed to the K2 relay contact failure, again a state-of-component fault. We can see, therefore, that a primary

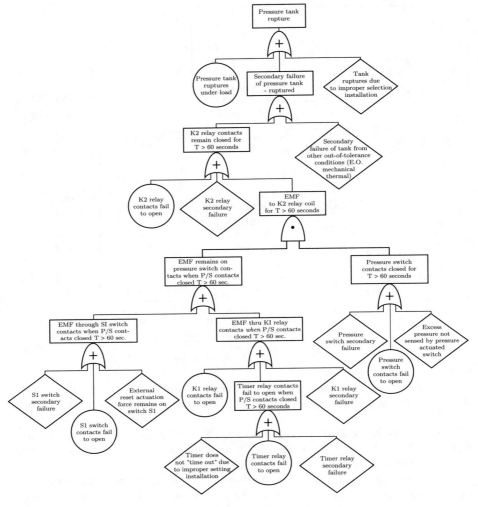

FIG. 7.2.5. *Pressure tank fault tree.*

failure of the K2 contacts could lead to system failure, a very uncomfortable situation. We should propose design modifications to include the timer relay contacts in the pump motor circuit, thus eliminating a one-event system failure. Note, too, that safety devices should directly monitor the system and not monitor other safety devices as does the timer relay in this example.

Exercises

7.2.1. Using the information on the hot water tank, construct a fault tree similar to Figure 7.2.3 for the top event: *failure to supply heated water.*

7.2.2. For the simple battery/motor diagram below, construct a fault tree for the top event: *motor does not start*. Assume the wires and connections are perfect. Consider two additional components of the motor: (1) the brushes and (2) the coil. Initially both switches are open.

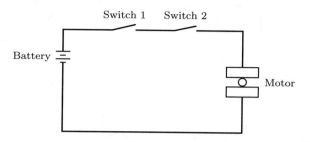

7.2.3. Redraw the pressure tank fault tree, Figure 7.2.5, with the following design modification. Put the timer contacts in both the control circuit and the pump circuit.

7.2.4. Represent an EXCLUSIVE OR gate as a fault tree using only OR, AND, and NOT gates. Make a "truth table" for an EXCLUSIVE OR gate and see if the tree constructed satisfies the truth table. The following is a truth table for an INCLUSIVE OR gate.

Input event A	Input event B	Output event
T	T	T
T	F	T
F	T	T
F	F	F

7.3. Fault tree probability analysis.

In the fault tree of Figure 7.2.3, there are 15 basic and secondary events. If we assign probabilities to these events, we can calculate the probability of the top event by simply starting at the bottom of the tree and working our way up assigning probabilities to the gate events in succession. This is because, in this case, there are no replicated events in the tree, i.e., events that appear in more than one place and also because there are *no NOT* gates. This simple calculation is not possible in general. We will illustrate algorithms for computing fault tree top event probability using the following abstract fault tree (see Figure 7.3.1). Note that event 6 is a replicated event since it appears twice in the tree.

A method for obtaining minimal cut sets (MOCUS).   A set of basic fault events constitutes a cut set if their occurrence causes the top event to occur. We

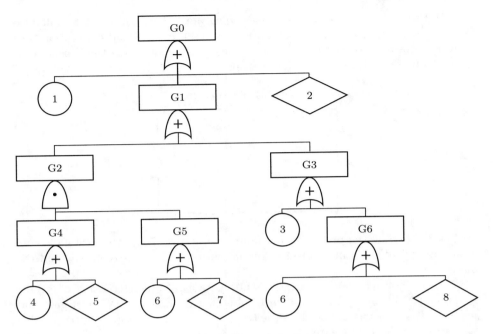

FIG. 7.3.1. *Example fault tree.*

use the term "cut set" since the top event is usually a failure event even though the term "path set" may seem more natural. The set is minimal if it cannot be reduced and still cause the top event to occur. We can calculate the probability of the top event using the minimal cut sets and the probabilities of primary events and undeveloped events.

An algorithm called MOCUS (method for obtaining cut sets) starts at the top and works its way down. Using Figure 7.3.1 to illustrate, start with the top event. Since the gate below it is an OR gate, list immediate input events in separate rows as follows:

1

2

G1

Since G1 is an OR gate, replace G1 by two additional rows as follows:

1

2

G2

G3

Since G2 is an AND gate, put its input events in the same row, obtaining

$$1$$
$$2$$
$$\text{G4, \ G5}$$
$$\text{G3}$$

We continue in this way, always
    (1) listing input events in separate rows if the gate is an OR gate, and
    (2) listing input events in the same row if the gate is an AND gate.
Finally, we obtain

$$1$$
$$2$$
$$4, 6$$
$$4, 7$$
$$5, 6$$
$$5, 7$$
$$3$$
$$6$$
$$8$$

These are the Boolean indicated cut sets (BICS). However, they are not the minimal cut sets since basic event 6 appears in three rows. We discard the supersets containing event 6 since they are not minimal, leaving the family of minimal cut sets:

$$[\{1\}, \ \{2\}, \{4, 7\}, \{5, 7\}, \{3\}, \{6\}, \{8\}]. \tag{7.3.1}$$

Note that there are nine BICS but only seven minimal cut sets. The time required to find and discard supersets is the only drawback to this algorithm.

Dual fault trees and BICS. The dual tree to Figure 7.3.1 can be found by replacing OR gates by AND gates and AND gates by OR gates. Basic events and gate events are replaced by their negation. The minimal cut sets of either the primal tree or its dual can be used to calculate the probability of the top event. Sometimes it is advantageous to use the dual tree. We can quickly determine the number of BICS for both the primal tree and its dual and use this information to determine which tree to use to find the minimal cut sets.

Calculating the number of BICS. To calculate the number of BICS, first assign weight 1 to each basic event in the tree. Starting at the bottom of the tree, assign a weight to each gate event as follows:
    (1) for each OR gate assign the SUM of the input weights;
    (2) for each AND gate assign the PRODUCT of the input weights.
    The final weight assigned to the top event is the number of BICS. This is a very fast algorithm.

Calculating the maximum size of BICS. We can also determine the maximum number of events in a BICS as follows. As before assign weight 1 to each basic event in the tree. Starting at the bottom of the tree,

(1) for each OR gate assign the *MAXIMUM* of the input weights;

(2) for each AND gate assign the *SUM* of the input weights.

The final weight assigned to the top event is the maximum number of events in a BICS. This is a very fast algorithm.

Calculating the probability of the top event. After assigning probabilities to the primary and secondary basic events, the problem is to calculate the probability of the top event. If there are no replicated events or NOT gates, this is very easy. Starting with the basic events, calculate the probability of each gate event having basic events as inputs. Continue in this manner until the top event is reached.

If there are no NOT gates but there are replicated basic events, then we find the minimal cuts for either the primal tree or the dual tree, whichever has the least number of BICS. With the minimal cuts in hand, we can develop the minimal path Boolean representation for the top event as in section 6.2.

Note that although we use the term minimal cuts for the fault tree, they are really like path sets relative to the occurrence of the top event (usually a failure event). This confusion is due to the success orientation of the network as opposed to the failure orientation of the fault tree.

With the Boolean representation for the top event we can use Boolean probability reduction to compute the probability of the top event. Another approach would be to use the principle of inclusion–exclusion with respect to the minimal cut events as in section 6.2. Again, the minimal cuts are interpreted as minimal paths as the way in which the principle of inclusion–exclusion is used in Chapter 6.

Boolean reduction. If the fault tree has no NOT gates we can use the method of Boolean reduction to calculate the probability of the top event. Let $K_1, K_2, \ldots, K_k$ be the k minimal cut sets for a given fault tree. Then

$$\psi(\mathbf{x}) = \coprod_{j=1}^{k} \prod_{i \in K_j} x_i. \qquad (7.3.2)$$

The structure function ψ in this case is 1 if the top event occurs (a failure event) and 0 otherwise. This was called the minimal path representation in Chapter 6, but because we are now failure oriented it is now the minimal cut representation for fault trees.

Substituting random quantities X_i for x_i and taking the expectation we obtain

$$\Pr(\text{top event}) = \Pr[\psi(\mathbf{X}) = 1] = E[\psi(\mathbf{X})].$$

The principle of inclusion–exclusion. If the fault tree has no NOT gates we can use the principle of inclusion–exclusion to calculate the probability of the top event. Let E_i be the event in which all basic events in K_i occur. Then by the principle of inclusion–exclusion we have

$$\text{Pr(top event)} = \sum_{i=1}^{k} \text{Pr}[E_i] - \sum_{i<j} \text{Pr}[E_i E_j] + \cdots + (-1)^{k-1}\text{Pr}[E_1 E_2 \ldots E_k].$$

Exercises

7.3.1. Use the following probabilities of failure per hot water tank use for events in the fault tree in Figure 7.2.3 to calculate the probability of the top event. The probability for tank rupture is 10^{-6} per use. All other primary basic events (denoted by circles) have probability 10^{-4} per use. Diamond events or undeveloped events have probability 10^{-3} per use.

7.3.2. For the pressure tank fault tree, Figure 7.2.5, find the minimal cuts using the MOCUS algorithm. Calculate the probability of the top event assuming the following probabilities. Pressure tank rupture is 10^{-8} per use while the event that the pressure switch contacts are jammed closed is 10^{-4}. All other primary (circle) events are 10^{-5} per use. Assume that secondary diamond events have probability 0 per use.

7.3.3. (a) Apply both the number of the BICS algorithm and the maximum size of the BICS algorithm to Figure 7.3.1.

(b) Give short proofs of both the number of the BICS algorithm and the maximum size of the BICS algorithm.

7.3.4. Give the Boolean representation for the structure function corresponding to an EXCLUSIVE OR gate.

7.4.   Notes and references.

Section 7.1. One of the best references for fault tree construction is the 1981 "Fault Tree Handbook" available from the U.S. Nuclear Regulatory Commission as NUREG-0492 (see Vesely et al. (1981)).

Section 7.2. The fault tree construction approach is due to David Haasl. Common mistakes in fault tree construction are

(1) to omit the system inductive analysis step that should include a table listing critical components' failure modes and effects and

(2) to ignore the first rule of fault tree construction, namely, to determine the immediate necessary and sufficient causes for a gate event.

Section 7.3. MOCUS and the BICS algorithms are due to Fussell and Vesely (1972).

CHAPTER 8

System Availability and Maintainability

In this chapter we assume some familiarity with *renewal theory* although most of the formulas used will be intuitive. (See Barlow and Proschan (1996), Chapter 3, section 2.)

8.1. Introduction.

Since many systems are subject to inspection and repair policies, availability is often the main reliability measure of interest. In sections 8.3 and 8.4 we discuss methodology for this calculation.

As in Chapter 6, we start with a network representation of a system of interest. In place of unreliable arcs we now refer to unreliable components. Again we assume that component failure events are *statistically independent*. However, now component indicator random variables will depend on time as the following notation suggests. We refer to *component positions* since with repair there may be many components used in a particular position. Let

$$X_i(t) = \begin{cases} 1 & \text{if the component in position } i \text{ is working at time } t, \\ 0 & \text{otherwise} \end{cases} \qquad (8.1.1)$$

for $i = 1, 2, \ldots, n$.

Let ϕ be a *coherent* system structure (organizing) function as in Chapter 6 and let

$$\phi(X_1(t), X_2(t), \ldots, X_n(t)) = \begin{cases} 1 & \text{if the system is working at time } t, \\ 0 & \text{otherwise.} \end{cases}$$
$$(8.1.2)$$

Availability at time t, $A(t)$, is defined as

$$A(t) = P\left[\phi(X_1(t), X_2(t), \ldots, X_n(t)) = 1\right] = E\left[\phi(X_1(t), X_2(t), \ldots, X_n(t))\right],$$

i.e., the probability that the system is working at time t. With repair possible, the system could have failed before time t but could also have been repaired before time t. Even assuming component failure events are independent, $A(t)$ is in general difficult to calculate. For this reason we will obtain asymptotic $(t \uparrow \infty)$ results, which should provide good approximations for long time intervals or, at the very least, some rule-of-thumb measures.

Consider a *single* component or a system considered as a single component having mean life μ and mean repair time v. In this case, where component uptimes are independent and also component downtimes are independent, the limiting availability is easy to compute. We have

$$A(t) \to \frac{\mu}{\mu + v} \qquad (8.1.3)$$

as $t \to \infty$. This is intuitively true since, roughly, the probability that the system is up will be the mean uptime divided by the mean length of a cycle consisting of an uptime and a downtime.

When we consider a complex system with many component positions with different component lifetimes and component repair times, it is not easy to calculate system availability.

8.2. Calculating system reliability in the absence of repair.

Let T_i be the random time to failure of a component in position i and $F_i(t) = P(T_i \le t)$. Components used in position i are considered exchangeable. We will also assume that

$$\phi(X_1(0), X_2(0), \ldots, X_n(0)) = 1;$$

i.e., the system is initially up.

Let

$$T = \text{minimum} \{t : \phi(X_1(t), X_2(t), \ldots, X_n(t)) = 0\}. \qquad (8.2.1)$$

Then, without the possibility of repair, which we assume at this point, T is the time to system failure.

Given $F_i(t)$, $i = 1, 2, \ldots, n$, our first problem is to compute $\overline{F}(t) = P(T > t)$, the probability that the system survives t time units in the absence of any component repair.

Fix t for the moment and let

$$p_i = P(T_i > t) = \overline{F}_i(t),$$

where p_i stands for the probability that the component in the ith position is working at time t (or the arc reliability at time t as in Chapter 6). We claim that network system reliability at time t is, in our new notation, the same as the probability that the system survives to time t so that

$$P(T > t) = h_\phi(p_1, p_2, \ldots, p_n). \qquad (8.2.2)$$

This is so since the system is coherent and thus fails as soon as the first network minimal cut set fails (in the sense that all components in some minimal cut set fail first before all components fail in any other minimal cut set).

To compute $\overline{F}(t) = P(T > t)$, we need only employ a suitable network reliability algorithm from Chapter 6 to compute the system reliability at time

t, replacing p_i by $\overline{F}_i(t) = P(T_i > t)$ in $h_\phi(p_1, p_2, \ldots, p_n)$. This can be easily graphed for selected values of $t \geq 0$.

To compute $E(T)$, it is sufficient to compute

$$\int_0^\infty \overline{F}(t)dt = E(T). \tag{8.2.3}$$

This is so since if $\int_0^\infty x dF(x) < \infty$, then $\lim_{t \to \infty} \overline{F}(t) = 0$ and $\int_0^\infty x dF(x) = \int_0^\infty \overline{F}(x)dx$, using integration by parts.

Series systems. Consider a series system with n component positions. Let failure rate functions be $r_i(t)$ for components in position i, $t \geq 0$ and $i = 1, 2, \ldots, n$. Then the system survival probability without repair and in the case of independent failure events is

$$\overline{F}(t) = P(T > t) = \prod_{i=1}^n e^{-\int_0^t r_i(u)du}$$

$$= e^{-\int_0^t [\sum_{i=1}^n r_i(u)]du}. \tag{8.2.4}$$

In this case, the system failure rate function is $\Lambda(t) = \sum_{i=1}^n r_i(t)$.

If failure rate functions are constant, i.e.,

$$r_i(t) = \frac{1}{\mu_i} \quad \text{for } t \geq 0 \quad \text{and} \quad i = 1, 2, \ldots, n,$$

then the system failure rate function is also constant, say, Λ, and

$$\Lambda = \sum_{i=1}^n \frac{1}{\mu_i}. \tag{8.2.5}$$

The mean time to system failure μ in this series case is

$$\mu = \frac{1}{\sum_{i=1}^n \frac{1}{\mu_i}}. \tag{8.2.6}$$

Exercises

8.2.1. Use the reliability function for the bridge from equation (6.1.3), namely,

$$h(p) = 2p^5 - 5p^4 + 2p^3 + 2p^2,$$

to graph the bridge survival probability when $p_i = \overline{F}_i(t) = e^{-t/\theta}$ for $t \geq 0$, $i = 1, 2, \ldots, 5$, and $\theta = 10$ hours. On the same graph display the survival distribution for the exponential survival distribution, $e^{-t/\theta}$ for $t \geq 0$ and $\theta = 10$. Do the two survival distributions cross? If so, where?

Also compute the mean time to system failure. Is it less than or greater than $\theta = 10$ hours?

8.2.2. Use the minimal path sets in section 6.2,

$$\mathbf{P} = [\{2,5\}, \{1,3,5\}, \{2,e,4\}, \{1,3,4\}, \{1,2,4\}],$$

to graph the system survival distribution for the rooted directed network in Figure 6.2.3 when

$$p_i = \overline{F}_i(t) = e^{-\frac{t}{\theta}} \quad \text{for } t \geq 0, \quad i = 1, 2, \ldots, 6$$

and $\theta = 10$ hours. Assume independent failure events.

8.2.3. Graph the system survival distribution for the hot water tank fault tree in Figure 7.2.3. Use the following exponential failure rate for events in the fault tree. The failure rate for tank rupture is 10^{-6} per use. All other primary basic events (denoted by circles) have failure rate 10^{-4} per use. Diamond events or undeveloped events have failure rate 10^{-3} per use. Consider time continuous, although it is really in the number of hot water tank usages.

8.3. Coherent systems with separately maintained components.

Let components used in position i have mean lifetime μ_i with mean repair time v_i. Furthermore, suppose that components are repaired immediately at failure while other nonfailed components continue to operate or at least remain "turned on." In this context, the system fails only when a minimal cut set of components requires repair. If the system is a series system, it fails as soon as any one component fails. However, even with one component under repair, other working components continue to operate and may fail during the repair of the original failed component. Because of these assumptions, system availability at time t for an arbitrary coherent structure ϕ will be

$$A(t) = P\left[\phi(\mathbf{X}(t)) = 1\right] = h\left[A_1(t), A_2(t), \ldots, A_n(t)\right], \qquad (8.3.1)$$

where $\mathbf{X}(t) = (X_1(t), X_2(t), \ldots, X_n(t))$ and $A_i(t)$ is the availability of the component in position i at time t. Assume that we have a renewal process in each component position; i.e., the sequence of failure times and repair times in a given component position are independent and identically distributed. It follows that for component position i,

$$\lim_{t \to \infty} A_i(t) = \frac{\mu_i}{\mu_i + v_i} \overset{DEF}{=} A_i.$$

Since components in position i have mean lifetime μ_i and mean repair time v_i and since components are separately maintained, system availability is, in the limit as $t \to \infty$,

$$A(t) \to h\left[\frac{\mu_1}{\mu_1 + v_1}, \frac{\mu_2}{\mu_2 + v_2}, \ldots, \frac{\mu_n}{\mu_n + v_n}\right]. \qquad (8.3.2)$$

This can be computed using the network algorithms for calculating $h(\mathbf{p})$ in Chapter 6.

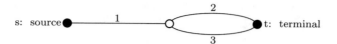

FIG. 8.3.1. *Network with three component (arc) positions.*

Example. For the network in Figure 8.3.1, system availability at time t is

$$A(t) = A_1(t) \left[A_2(t) \coprod A_3(t) \right]$$

and

$$A = \lim_{t \to \infty} A(t) = A_1 \left[A_2 \coprod A_3 \right] = \frac{\mu_1}{\mu_1 + v_1} \left[\frac{\mu_2}{\mu_2 + v_2} \coprod \frac{\mu_3}{\mu_3 + v_3} \right].$$

We may ask many more questions concerning system performance in the case of separately maintained components in coherent systems:

(1) How often do components in position i fail and "cause" the system to fail? By "cause" we mean that the component in position i fails and the system then fails due to component i failure.

(2) What is the long run (asymptotic as $t \to \infty$) system failure rate, i.e., the expected frequency of system failures per unit time?

(3) What is the long run average system up- (down) time?

In each component position we assume that we have a renewal counting process. The renewal counting process in component position i will be $\{N_i(t); t \geq 0\}$. Let $\tilde{N}_i(t)$ be the number of times in $[0, t]$ that components in position i fail and *cause* system failure in $[0, t]$. We will show that, asymptotically, the expected frequency with which components in position i fail and cause system failure is

$$\lim_{t \to \infty} \frac{E[\tilde{N}_i(t)]}{t} = \frac{h(1_i, \mathbf{A}) - h(0_i, \mathbf{A})}{\mu_i + v_i}, \tag{8.3.3}$$

where $(1_i, \mathbf{A}) = (A_1, A_2, \ldots, 1_i, \ldots, A_n)$ and $(0_i, \mathbf{A}) = (A_1, A_2, \ldots, 0_i, \ldots, A_n)$.

Proof of formula (8.3.3). To show (8.3.3) it is sufficient to note that since the system is assumed coherent

$$\phi(1_i, \mathbf{x}(t)) - \phi(0_i, \mathbf{x}(t)) = \begin{cases} 1, \\ 0. \end{cases}$$

If $\phi(1_i, \mathbf{x}(t))$ and $\phi(0_i, \mathbf{x}(t))$ are both 1 or both 0, the difference is 0. The difference is 1 if and only if $\phi(1_i, \mathbf{x}(t)) = 1$ and $\phi(0_i, \mathbf{x}(t)) = 0$. By coherence we cannot have $\phi(1_i, \mathbf{x}(t)) = 0$ and $\phi(0_i, \mathbf{x}(t)) = 1$. We say that the component in position i is *critical at time t* if

$$\phi(1_i, \mathbf{x}(t)) - \phi(0_i, \mathbf{x}(t)) = 1.$$

This follows because if the component in position i were to fail in the next instant after time t, the system would also fail.

Now let $N_i(t)$ be the cumulative number of renewals of components in position i by time t. Partition the time interval $[0, t]$ as follows:

For Δ_j sufficiently small and F_i continuous, the number of failures in component position i during the time interval Δ_j will be either 0 or 1; i.e.,

$$N_i(\Delta_j) = \begin{cases} 1 & \text{if a failure occurs in } \Delta_j, \\ 0 & \text{otherwise.} \end{cases}$$

Hence

$$\tilde{N}_i(t) \cong \sum_{j=1}^{k} [\phi(1_i, X(u_j)) - \phi(0_i, X(u_j))] N_i(\Delta_j),$$

where $u_j \in \Delta_j$, since the component in position i causes system failure only if i fails and at the time of failure it is critical to the system.

Thus, using independence of events in different component positions we have

$$E[\tilde{N}_i(t)] \cong \sum_{j=1}^{k} E[\phi(1_i, X(u_j)) - \phi(0_i, X(u_j))] E[N_i(\Delta_j)].$$

Let $M_i(t) = E[N_i(t)]$, the renewal function for components in position i. Again, approximately

$$\frac{dM_i(u)}{du}$$

is the probability that the component in position i fails in the interval $[u, d+du)$. Hence, as $k \uparrow \infty$,

$$E[\tilde{N}_i(t)] \cong \int_0^t [h(1_i, \mathbf{A}(u)) - h(0_i, \mathbf{A}(u))] \, dM_i(u).$$

Since $A_i(t) \to \frac{\mu_i}{\mu_i + v_i}$ and for fixed T (very large) with $t > T$,

$$\frac{E[\tilde{N}_i(t)]}{t} \cong \frac{1}{t} \int_0^T [h(1_i, \mathbf{A}(u)) - h(0_i, \mathbf{A}(u))] \, dM_i(u)$$
$$+ \frac{1}{t} \int_T^t [h(1_i, \mathbf{A}(u)) - h(0_i, \mathbf{A}(u))] \, dM_i(u).$$

Since T is fixed, the first term goes to 0 as $t \uparrow \infty$. For large u,

$$[h(1_i, \mathbf{A}(u)) - h(0_i, \mathbf{A}(u))] \cong [h(1_i, \mathbf{A}) - h(0_i, \mathbf{A})]$$

and

$$\lim_{t\uparrow\infty} \frac{E[\tilde{N}_i(t)]}{t} = \frac{[h(1_i, \mathbf{A}) - h(0_i, \mathbf{A})]}{u_i + v_i}. \tag{8.3.4}$$

We have used the elementary renewal result, namely,

$$\lim_{t\to\infty} \frac{M(t)}{t} = \frac{1}{\mu_i + v_i}$$

to prove (8.3.4).   □

Let $\tilde{N}(t) = \sum_{i=1}^{n} \tilde{N}_i(t)$ be the number of system failures in $[0, t]$. It follows that the long run expected *frequency of system failures* is

$$\lim_{t\to\infty} \frac{E[\tilde{N}(t)]}{t} = \lim_{t\to\infty} \frac{\sum_{i=1}^{n} E[\tilde{N}_i(t)]}{t} = \sum_{i=1}^{n} \frac{h(1_i, \mathbf{A}) - h(0_i, \mathbf{A})}{\mu_i + v_i} \tag{8.3.5}$$

by equation (8.3.3).

System mean time between failures (MTBF).   Let $U_1, U_2, \ldots, U_k$ be successive system uptimes and $D_1, D_2, \ldots, D_k$ be successive system downtimes. From (8.3.5) we can deduce the long run MTBF by considering the first k uptime–downtime cycles and then letting $k \uparrow \infty$, namely,

$$\lim_{k\to\infty} \frac{U_1 + U_2 + \cdots + U_k + D_1 + D_2 + \cdots + D_k}{k} = \frac{1}{\sum_{i=1}^{n} \frac{[h(1_i,\mathbf{A})-h(0_i,\mathbf{A})]}{\mu_i+v_i}}.$$

System uptimes and downtimes.   At any one time the system may be down due to more than one component. Hence we cannot simply add up component downtimes contributing to system failure.

Since $U_1, U_2, \ldots, U_k$ are successive system uptimes and $D_1, D_2, \ldots, D_k$ are successive system downtimes, then it can be shown that

$$\frac{U_1 + U_2 + \cdots + U_k}{U_1 + U_2 + \cdots U_k + D_1 + D_2 + \cdots + D_k} \underset{k\to\infty}{\to} h[\mathbf{A}]. \tag{8.3.6}$$

This is true by the strong law of large numbers and since we assume $\lim_{t\to\infty} h[\mathbf{A}(t)]$ exists.

Since

$$\lim_{k\to\infty} \frac{k}{U_1 + U_2 + \cdots + U_k + D_1 + D_2 + \cdots + D_k}$$
$$= \lim_{t\to\infty} \frac{\tilde{N}(t)}{t} = \sum_{i=1}^{n} \frac{[h(1_i, \mathbf{A}) - h(0_i, \mathbf{A})]}{\mu_i + v_i} \tag{8.3.7}$$

by the strong law of large numbers we have

$$\lim_{k\to\infty} \frac{U_1 + U_2 + \cdots + U_k}{k} = \frac{h[\mathbf{A}]}{\sum_{i=1}^{n} \frac{[h(1_i,\mathbf{A})-h(0_i,\mathbf{A})]}{\mu_i+v_i}}. \tag{8.3.8}$$

Also, the long run average of system downtimes can be similarly calculated, namely,

$$\lim_{k \to \infty} \frac{D_1 + D_2 + \cdots + D_k}{k} = \frac{1 - h[\mathbf{A}]}{\sum_{i=1}^{n} \frac{[h(1_i, \mathbf{A}) - h(0_i, \mathbf{A})]}{\mu_i + v_i}}. \tag{8.3.9}$$

Example 8.3.1.   Series systems with separately maintained components.   In the case of a series system, any component failure will cause system failure. The long run system availability is

$$\lim_{t \uparrow \infty} A(t) = \prod_{j=1}^{n} \frac{\mu_j}{\mu_j + v_j} = A, \tag{8.3.10}$$

while the long run series system expected frequency of system failures from (8.3.5) is

$$\lim_{t \to \infty} \frac{\tilde{N}(t)}{t} = \sum_{i=1}^{n} \left[\frac{h(1_i, \mathbf{A}) - h(0_i, \mathbf{A})}{\mu_i + v_i} \right]$$

$$= \sum_{i=1}^{n} \left[\frac{1}{\mu_i + v_i} \right] \left[\prod_{j \neq i} \frac{\mu_j}{\mu_j + v_j} \right] \tag{8.3.11}$$

$$= \left[\sum_{j=1}^{n} \frac{\mu_j}{\mu_j + v_j} \right] \sum_{i=1}^{n} \frac{1}{\mu_i} = A \sum_{i=1}^{n} \frac{1}{\mu_i}.$$

The asymptotic MTBF is

$$\frac{1}{A \sum_{i=1}^{n} \frac{1}{\mu_i}}.$$

From (8.3.8) the long run average system uptime is

$$\lim_{k \to \infty} \frac{U_1 + U_2 + \cdots + U_k}{k} = \frac{1}{\sum_{i=1}^{n} \frac{1}{\mu_i}}. \tag{8.3.12}$$

It is interesting that in the series case with separately maintained components, the average of system uptimes does *not* depend on mean component downtimes.

The long run average of system downtimes, however, does depend on both mean component downtimes v_i and mean component uptimes μ_i; i.e.,

$$\lim_{k \to \infty} \frac{D_1 + D_2, + \cdots + D_k}{k} = \frac{1 - \prod_{j=1}^{n} \frac{\mu_j}{\mu_j + v_j}}{\left[\prod_{j=1}^{n} \frac{\mu_j}{\mu_j + v_j} \right] \sum_{i=1}^{n} \frac{1}{\mu_i}} = \frac{1 - A}{A \sum_{i=1}^{n} \frac{1}{\mu_i}}. \tag{8.3.13}$$

Exercises

8.3.1. For the network in Figure 8.3.1, calculate $\lim_{t \to \infty} \frac{E[\tilde{N}_i(t)]}{t}$ for each of the three components in the figure. Use these to calculate $\lim_{t \to \infty} \frac{E[\tilde{N}(t)]}{t}$. What

FIG. 8.4.1. *Each component failure shuts down all components.*

FIG. 8.4.2. *Components separately maintained.*

FIG. 8.4.3. *Component 1 shuts down 2 and 3 (not vice versa).*

is the MTBF? What is the long run average of system downtimes?

8.3.2. Parallel Systems. Let components in position i have mean lifetime μ_i with mean repair time v_i. Furthermore, suppose that components are repaired immediately at failure while other nonfailed components continue to operate or at least remain "turned on" and can fail. Just as we did for the series system, for a parallel system with n *separately maintained components* find the corresponding formulas for

(1) the asymptotic expected system failure rate,
(2) the asymptotic MTBF,
(3) the asymptotic average system uptime,
(4) the asymptotic average system downtime.

8.4. Suspended animation for components in series systems

In many cases, systems under consideration are essentially series systems. Although this implies that failure of any component causes system failure, it may or may not imply that a component failure results in other system components being shut off. A component that is shut down (but not because of failure) will be assumed to be in *suspended animation*; that is, it cannot fail in this state but also it cannot age in this state.

We illustrate the possible shut-off policies for series systems by the following diagrams. In Figure 8.4.1 the double-headed arrows indicate, for example, that component 1 shuts down component 2, which in turn shuts down component 3, etc. and vice versa. If components are separately maintained, there are no arrows at all. See Figure 8.4.2. If, for example, component 1 shuts down components 2 and 3, but not vice versa, we have Figure 8.4.3. Our task is to develop availability formulas for series systems with different shut-off policies. We have already done this in Example 8.3.1 in the case of separately maintained components.

Each component failure shuts down all other components. We first calculate appropriate formulas for the case where each component shuts down all others as in Figure 8.4.1. In each component position we assume that we have a renewal counting process. The renewal counting process in component position i will be $\{N_i(t); t \geq 0\}$. Let $U(t)$ for $t \geq 0$ be the total system uptime in the interval $[0, t]$.

THEOREM 8.4.1. *For series systems with independent component failure and repair events and the shut-off policy in Figure 8.4.1, we have the result*

$$\lim_{t \to \infty} \frac{U(t)}{t} = \left[1 + \sum_{i=1}^{n} \frac{v_i}{\mu_i}\right]^{-1} \tag{8.4.1}$$

almost surely. (Recall that shut-off components are in a state of suspended animation.) For a proof, see Barlow and Proschan (1973).

From (8.4.1) we can calculate the limiting average availability A_∞, namely,

$$A_\infty = \lim_{t \to \infty} E\left[\frac{U(t)}{t}\right] = \left[1 + \sum_{i=1}^{n} \frac{v_i}{\mu_i}\right]^{-1}. \tag{8.4.2}$$

Note that for $n = 1$,

$$A_\infty = \frac{1}{1 + \frac{v_1}{\mu_1}} = \frac{\mu_1}{\mu_1 + v_1}, \tag{8.4.3}$$

so that for $n = 1$ (8.4.2) agrees with the availability formula (8.1.3) given at the beginning of this chapter.

Let $D_i(t)$ be the total system downtime in $[0, t]$ due to failures of component i and $D_{i,\infty} = \lim_{t \to \infty} \frac{D_i(t)}{t}$. Let

$$\mu = \frac{1}{\sum_{i=1}^{n} \frac{1}{\mu_i}}$$

be the limiting average of series system uptimes [cf. (8.3.12)]. Using Theorem 8.4.1 we can prove the following corollary.

COROLLARY 8.4.2. *For series systems with independent component failure and repair events and the shut-off policy in Figure 8.4.1, the limiting average system downtime due to component i is*

$$D_{i,\infty} = \mu \frac{v_i}{\mu_i},$$

while the limiting average system downtime due to all component failure events is $D_\infty = \sum_{i=1}^{n} D_{i,\infty} = \mu \sum_{i=1}^{n} \frac{v_i}{\mu_i}$.

System MTBF. *MTBF* is the average of the sum of the successive system up- and downtimes. We already know that $\mu = [\sum_{i=1}^{n} \frac{1}{\mu_i}]^{-1}$ is the limiting average of series system uptimes and that $\mu \sum_{i=1}^{n} \frac{v_i}{\mu_i}$ is the limiting average of system

TABLE 8.4.1
Comparison of asymptotic maintainability measures.

	Separately maintained	Suspended animation
Availability	$h\left[\dfrac{\mu_1}{\mu_1+v_1}, \dfrac{\mu_2}{\mu_2+v_2}, \ldots, \dfrac{\mu_n}{\mu_n+v_n}\right]$	$\left[1 + \displaystyle\sum_{i=1}^{n} \dfrac{v_i}{\mu_i}\right]^{-1}$
Average uptime	$\dfrac{h[\mathbf{A}]}{\sum_{i=1}^{n} \frac{[h(1_i,\mathbf{A})-h(0_i,\mathbf{A})]}{\mu_i+v_i}}$	$\mu = \left[\displaystyle\sum_{i=1}^{n} \dfrac{1}{\mu_i}\right]^{-1}$
Expected failure frequency	$\displaystyle\sum_{i=1}^{n} \dfrac{h(1_i,A)-h(0_i,A)}{\mu_i+v_i}$	$\dfrac{1}{\mu + \mu \sum_{i=1}^{n} \frac{v_i}{\mu_i}}$
MTBF	$\dfrac{1}{\sum_{i=1}^{n} \frac{[h(1_i,\mathbf{A})-h(0_i,\mathbf{A})]}{\mu_i+v_i}}$	$\mu + \mu \displaystyle\sum_{i=1}^{n} \dfrac{v_i}{\mu_i}$
Average downtime	$\dfrac{1-h[\mathbf{A}]}{\sum_{i=1}^{n} \frac{[h(1_i,\mathbf{A})-h(0_i,\mathbf{A})]}{\mu_i+v_i}}$	$\mu \displaystyle\sum_{i=1}^{n} \dfrac{v_i}{\mu_i}$

downtimes. Therefore, it is not surprising that

$$\mathbf{MTBF} = \mu + \mu \sum_{i=1}^{n} \frac{v_i}{\mu_i}.$$

The limiting expected system failure frequency Λ is $\frac{1}{\mathbf{MTBF}}$ or

$$\Lambda = \frac{1}{\mu + \mu \sum_{i=1}^{n} \frac{v_i}{\mu_i}}.$$

see Table 8.4.1.

Exercises

8.4.1. Verify that when $n = 1$, all formulas for systems with separately maintained components and systems with suspended animation components agree.

8.4.2. Compare system availability when $\frac{v_i}{\mu_i} \approx$ constant for $i = 1, 2, \ldots, n$ using the formula for separately maintained components versus the formula for the case when shut-off components are in suspended animation.

8.4.3* (this exercise presumes knowledge of Markov processes). Suppose components 1 and 2 are in series. Component 1 shuts down component 2 but not vice versa. When component 1 fails and shuts down component 2, component 2 is in suspended animation and cannot fail. On the other hand, if component 2 fails, component 1 continues to operate and may fail while component 2 is being repaired. Repair commences immediately upon failure of either or both components. The diagram is as follows:

$$1 \qquad\qquad 2$$

Assume all failure and repair distributions are exponential with failure rates λ_i, θ_i for components $i = 1, 2$. *Remember that exponential distributions have the memoryless property.*

(a) Identify all possible system states. Let 0 denote the state when both components are working. Draw a sample space diagram of a possible scenario.

(b) Draw a state space diagram with transition rates indicated.

(c) If $\{Z(t), t \geq 0\}$ is the system state process, how would you calculate

$$\lim_{t \to \infty} P(Z(t) = 0) = \pi_0?$$

8.5.   Notes and references.

Section 8.1. Availability theory as presented in this chapter was partly developed in a paper by Barlow and Proschan (1973).

Section 8.2. The technique for calculating the distribution of time to system failure requires that the system be coherent. This is because only in this case can we be assured that the system reliability function is an increasing function of component reliabilities.

Section 8.3. The asymptotic formulas for separately maintained components were first developed by Ross (1975). The proof of formula (8.3.3) was developed in Barlow and Proschan (1975).

Section 8.4. Suspended animation for components in series systems was first considered by Bazovsky et al. (1962). However, they only considered the exponential case. The development in this section follows that in Barlow and Proschan (1973).

Notation

$A(t)$	system availability at time t
Λ	asymptotic expected system failure rate
MTBF	system mean time between failures
μ	asymptotic system MTBF

CHAPTER 9

Influence Diagrams

9.1. Using influence diagrams.

Mathematically, fault trees are probabilistic influence diagrams. However, the fault tree approach was based on engineering considerations and was invented by mechanical engineers. Fault trees are based on principles similar to the reasoning used by engineers.

Influence diagrams, on the other hand, were invented by decision analysts as an alternative to decision trees. Influence diagrams provide an excellent graphical tool for understanding probabilistic conditional independence. The influence diagram is a graphical representation of the relationships between random quantities that are judged relevant to a real problem.

For analyzing system failure events, the fault tree is the more useful representation. For decision problems, and particularly those involving monetary considerations, the influence diagram is the more useful representation. In the following sections we provide a formal introduction to influence diagrams, emphasizing their probabilistic basis.

9.2. Probabilistic influence diagrams.

Definitions and basic results. An influence diagram is, first of all, a directed graph. A graph is a set V of nodes or vertices together with a set A of arcs joining the nodes. It is said to be directed if the arcs are arrows (directed arcs). Let $V = \{v_1, v_2, \ldots, v_n\}$ and let A be a set of ordered pairs of elements of V, representing the directed arcs. That is, if $[v_i, v_j] \in A$ for $1 \leq i, j \leq n$, then there is a directed arc (arrow) from vertex v_i to vertex v_j (the arrow is directed from v_i to v_j). If $[v_i, v_j] \in A$, v_i is said to be an adjacent predecessor of v_j and v_j is said to be an adjacent successor of v_i. The direction of arcs is meant to denote influence (or possible dependence).

Circles (or ovals) represent random quantities which may, at some time, be observed and consequently may change to data. Circle nodes are called probabilistic nodes. Attached to each circle node is a conditional probability (density) function. This function is a function of the state of the node and also of the states of the adjacent predecessor nodes.

FIG. 9.2.1. *Influence diagram symbols.*

A double circle (or double oval) denotes a deterministic node, which is a node with only one possible state, given the states of the adjacent predecessor nodes; i.e., it denotes a deterministic function of all adjacent predecessors. Thus, to include the background information H in the graph, we would have to use a double circle around H. See Figure 9.2.1.

The following concepts formalize the ideas used in drawing the diagrams.

DEFINITION. *A directed graph is cyclic and is called a cyclic directed graph, if there exists a sequence of ordered pairs in A such that the initial and terminal vertices are identical; i.e., there exists an integer $k \leq n$ and a sequence of k arcs of the following type:*

$$[v_{i_1}, v_{i_2}], [v_{i_2}, v_{i_3}], \ldots, [v_{i_{k-1}}, v_{i_k}], [v_{i_k}, v_{i_1}].$$

DEFINITION. *An acyclic directed graph is a directed graph that is not cyclic.*

DEFINITION. *A root node is a node with no adjacent predecessors. A sink node is a node with no adjacent successors. Note that any acyclic directed graph must have at least one root node and one sink node.*

DEFINITION. *A probabilistic influence diagram is an acyclic directed graph in which*

(i) *circle nodes represent random quantities while double circle nodes represent deterministic functions,*

(ii) *directed arcs indicate possible dependence, and*

(iii) *attached to each node is a conditional probability function (or deterministic function) for the node that depends on the states of adjacent predecessor nodes.*

Figure 9.2.2 illustrates the fault tree OR gate using the influence diagram symbols. Random quantities X_1 and X_2 are random indicators taking only the values 1 or 0. The deterministic function δ depends on X_1 and X_2 where $\delta(X_1, X_2) = \text{maximum } (X_1, X_2)$.

Given a directed acyclic graph together with node conditional probabilities (i.e., a probabilistic influence diagram), there exists a unique joint probability function corresponding to the random quantities represented by the nodes of the graph. This is because a directed graph is acyclic if and only if there exists a list ordering of the nodes such that any successor of a node x in the graph follows node x in the list as well. Consequently, following the list ordering and taking the product of all node conditional probabilities we obtain the joint probability of the random quantities corresponding to the nodes in the graph. Note that in a cyclic graph the product of the conditional probability functions attached to the nodes would not determine the joint probability function.

The following basic result shows that the *absence of an arc* connecting two nodes in the influence diagram denotes the judgment that the unknown quanti-

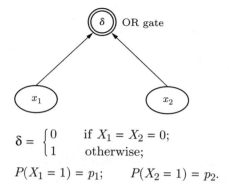

$$\delta = \begin{cases} 0 & \text{if } X_1 = X_2 = 0; \\ 1 & \text{otherwise;} \end{cases}$$

$$P(X_1 = 1) = p_1; \qquad P(X_2 = 1) = p_2.$$

FIG. 9.2.2.   *Influence diagram for fault tree OR gate.*

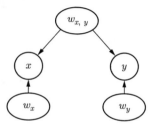

FIG. 9.2.3.   *x and y are conditionally independent.*

ties associated with these nodes are conditionally independent given the states of all adjacent predecessor nodes. For example, in Figure 9.2.2 root node random quantities X_1 and X_2 are independent, while the deterministic function δ depends on both X_1 and X_2.

Remark. Let x_i and x_j represent two nodes in a probabilistic influence diagram. If there is *no arc* connecting x_i and x_j, then x_i and x_j are conditionally independent given the states of the adjacent predecessor nodes; i.e.,

$$p(x_i, x_j \mid w_i, w_j, w_{ij}) = p(x_i \mid w_i, w_j, w_{ij})p(x_j \mid w_i, w_j, w_{ij}),$$

where $w_i(w_j)$ denotes the set of adjacent predecessor nodes to only $x_i(x_j)$, while w_{ij} denotes the set of adjacent predecessor nodes to both x_i and x_j.

Figure 9.2.3 illustrates conditional independence implied by the absence of an arc connecting x and y. In a probabilistic influence diagram, if two nodes, x_i and x_j, are root nodes then they are independent.

Exercises

9.2.1. In the following influence diagram, which nodes are root nodes? Which nodes are conditionally independent? Which nodes are independent? Which nodes are sink nodes?

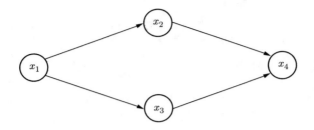

9.2.2. In the following influence diagram, which nodes are root nodes? Which nodes are conditionally independent? Which nodes are independent? Which nodes are sink nodes?

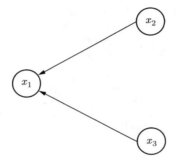

9.3. Probabilistic influence diagram operations.

The Bayesian approach to statistics is based on probability judgments and as such follows the laws of probability. You are said to be coherent if

(i) you use probability to measure your uncertainty about quantities of interest and

(ii) you do not violate the laws of probability when stating your uncertainty concerning probability assessments.

Probabilistic influence diagrams (and influence diagrams in general) are helpful in assuring coherence. Clearly, from coherence, any operation to be performed in a probabilistic influence diagram must not violate the laws of probability. The three basic probabilistic influence diagram operations that we discuss next are based on the addition and product laws. These operations are (1) splitting nodes, (2) merging nodes, and (3) arc reversal.

Splitting nodes. In general, a node in a probabilistic influence diagram can denote a vector random quantity. It is always possible to split such a node into other nodes corresponding to the elements of the vector random quantity. To illustrate ideas, suppose that a node corresponds to a vector of two random quantities x and y with joint probability function $p(x, y)$. From the product law of probability we know that

$$p(x, y) = p(x)p(y \mid x) = p(x \mid y)p(y).$$

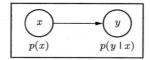 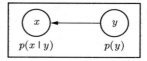

FIG. 9.3.1. *Equivalent probabilistic influence diagrams for two random quantities.*

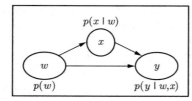 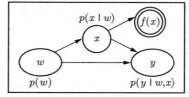

FIG. 9.3.2. *Addition of a deterministic node.*

Hence, Figure 9.3.1 presents the three possible probabilistic influence diagrams that can be used in this case showing the two ways of splitting node (x, y).

The following property is also a direct consequence of the laws of probability and it is of special interest for statistical applications.

Property. Let x be a random quantity represented by a node of a probabilistic influence diagram and let $f(x)$ be a (deterministic) function of x. Suppose we connect to the original diagram a deterministic node representing $f(x)$ using a directed arc from x to $f(x)$. Then the joint probability distributions for the two diagrams are equal. (See Figure 9.3.2 for illustration.)

Proof. Let w and y represent the sets of random quantities that precede and succeed x, respectively, in a list ordering. Note that $p(f(x) \mid w, x) = p(f(x) \mid x) = 1$ and consequently from the product law $p(x, f(x) \mid w) = p(x \mid w)$. That is, node x may be replaced by node $(x, f(x))$ without changing the joint probability of the graph nodes. Using the splitting node operation in node $(x, f(x))$ with x preceding $f(x)$, we obtain the original graph with the additional deterministic node $f(x)$ and a directed arc from x to $f(x)$. Note also that no other arc is necessary since $f(x)$ is determined by x and $p(y \mid w, f(x), x) = p(y \mid w, x)$. □

Merging nodes. The second probabilistic influence diagram operation is the merging of nodes. Consider first a probabilistic influence diagram with two nodes x and y with a directed arc from x to y. The product law states that $p(x, y) = p(x)p(y \mid x)$. Hence, without changing the joint probability of x and y, the original diagram can be replaced by a single node diagram representing the vector (x, y). The first two diagrams of Figure 9.3.1 in the reverse order illustrate this operation. In general, two nodes, x and y, can be replaced by a single node, representing the vector (x, y), if there is a list ordering such that x is an immediate predecessor or successor of y.

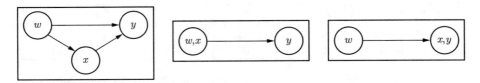

FIG. 9.3.3. *Diagram with adjacent nodes w and y not allowed to be merged.*

It is not always possible to merge two adjacent nodes in a probabilistic influence diagram. Note that two adjacent nodes may not be neighbors in any list ordering. For example, consider the first diagram of Figure 9.3.3. Note that all pairs of nodes in this diagram constitute adjacent nodes. However, w and y cannot be merged into a node representing (w, y). Clearly the only list ordering here is $w < x < y$ and w and y are not immediate neighbors in this ordering. The problem here is that to merge w and y we would need an arc from (w, y) to x and another from x to (w, y). The reason for this is the existence of arcs $[w, x]$ and $[x, y]$ in the original graph. If we were to have arcs in both directions between (w, y) and x, we would not obtain, in general, the joint probability function from the diagram since $p(w, x, y) \neq p(x \mid w, y) p(w, y \mid x)$. Also it can be seen from the first diagram of Figure 9.3.3 that there exist two paths from w to y. This is the graphical way to see that w and y cannot be merged into a single node. To construct a graphical technique to check if two nodes can be merged, we need the following definition and theorem.

DEFINITION. *A directed path from node x_i to node x_j is a chain of ordered pairs corresponding to directed arcs that lead from x_i to x_j.*

THEOREM 9.3.1 (merging nodes theorem). *In a probabilistic influence diagram, nodes x and y can be merged if either*

(1) *the only directed path between x and y is a directed arc connecting x and y or*

(2) *there is no directed path connecting x and y.*

Proof. To be definite, suppose that x precedes y in an associated list ordering corresponding to a probabilistic influence diagram. Let $w_x (w_y)$ be the set of adjacent predecessors of $x(y)$ but not of $y(x)$, and let w_{xy} be the set of nodes that are adjacent predecessors of both x and y. Since there is no directed path from x to y except, possibly, for a directed arc from x to y, we may add arcs from each node in w_x to y and from each node in w_y to x without creating any cycles. This is possible because directed arcs indicate possible dependence but not necessarily strict dependence. We have of course lost some graph information as a result of these arc additions.

In the associated list ordering of nodes for our modified diagram, the family of nodes $\{w_x, w_y, w_{xy}\}$ precedes both x and y. Since there is no other directed path from x to y other than possibly a directed arc from x to y, there exists an associated list ordering of nodes for which x is an immediate predecessor of y in this list ordering. The product

$$p(x \mid w_x, w_y, w_{xy}) p(y \mid x, w_x, w_{xy})$$

FIG. 9.3.4. *Reversing arc operation in a two node probabilistic influence diagram.*

must appear in the representation for the joint probability function for all probabilistic nodes based on the list ordering. Since

$$p(y, x \mid w_x, w_y, w_{xy}) = p(x \mid w_x, w_y, w_{xy})p(y \mid x, w_x, w_y, w_{xy})$$

by the product law, we can merge x and y.

Finally, suppose that there is a directed path from x to y other than a directed arc from x to y. In this case it is not difficult to see that merging x and y would create a cycle, which is not allowed. □

The above result is related to arc reversal, an important operation discussed next.

Reversing arcs.   The probabilistic influence diagram operation corresponding to Bayes' formula is that of arc reversal. Consider the diagram on the left in Figure 9.3.4. Using the merging nodes operation we obtain the single node diagram in the center where the probability function of the node (x, y) is obtained from the first diagram as $p(x, y) = p(x)p(y \mid x)$. Using the splitting nodes operation we can obtain the diagram on the right of Figure 9.3.4. Note that to obtain the corresponding probability functions we use

(1) the theorem of total probability for $p(y) = \sum_x p(y \mid x)p(x)$, where $\sum_x$ is the sum (or integral) over all possible values of x, and

(2) the multiplication law for $p(x \mid y) = \frac{p(x,y)}{p(x)}$ since $p(y)p(x \mid y) = p(x, y)$.

By substituting the appropriate expressions in $p(x \mid y)$ we obtain Bayes' formula. That is,

$$p(x \mid y) = \frac{p(x)p(y \mid x)}{\sum_x p(x)p(y \mid x)}.$$

Hence, by using the theorem of total probability and Bayes' formula when performing an arc reversal operation, we can go directly from the left diagram to the right one in Figure 9.3.4 without having to consider the one in the center.

Although the diagrams are different, they have the same joint probability function for node random quantities. This fact is formalized in the following definition.

DEFINITION. *Two probabilistic influence diagrams are said to be equivalent in probability if they have the same joint probability function for node random quantities.*

Consider the diagram of Figure 9.3.5 where w_x, w_y, and w_{xy} are sets of adjacent predecessors of x and (or) y as indicated by the figure. If arc $[x, y]$ is the only directed path from node x to node y, we may add arcs $[w_x, y]$ and $[w_y, x]$ to the diagram without introducing any cycles. (See left diagram of Figure 9.3.6.)

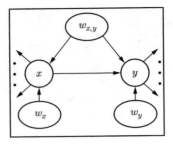

FIG. 9.3.5.

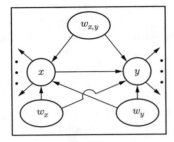
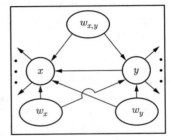

FIG. 9.3.6. *Equivalent probabilistic influence diagrams.*

Remember that a directed arc only indicates possible dependence. Probability nodes in the right diagram in Figure 9.3.6 are obtained from the left diagram by using Bayes' formula and the theorem of total probability.

The following result introduces the conditions under which arc reversal operations can be performed.

THEOREM 9.3.2 (reversing arcs theorem). *Suppose that arc $[x, y]$ connects nodes x and y in a probabilistic influence diagram. $[x, y]$ can be reversed to $[y, x]$ without changing the joint probability function of the diagram if*

(1) *there is no other directed path from x to y,*

(2) *all the adjacent predecessors of $x(y)$ in the original diagram become also adjacent predecessors of $y(x)$ in the modified diagram and*

(3) *the conditional probability functions attached to nodes x and y are also modified in accordance with the laws of probability.*

Proof. Let $w_x(w_y)$ be the set of adjacent predecessors of $x(y)$ but not of $y(x)$, and let w_{xy} be the set of adjacent predecessors of both x and y. Since arcs represent possible dependence, we can add arcs to the diagram in order to make the set (w_x, w_y, w_{xy}) an adjacent predecessor of both x and y. Since there is no other directed path connecting x and y, there is a list ordering such that x is an immediate predecessor of y in the list. Note also that the elements of the set (w_x, w_y, w_{xy}) are all predecessors of both x and y in the list ordering. To obtain the joint probability function corresponding to the first diagram, we consider the product, following the list ordering, of all node conditional probability functions.

As a factor of this product we have

$$p(x \mid w_x, w_y, w_{xy})p(y \mid x, w_y, w_{xy}) = p(x \mid w_x, w_y, w_{xy})p(y \mid x, w_x, w_y, w_{xy})$$
$$= p(x, y \mid w_x, w_y, w_{xy}) = p(y \mid w_x, w_y, w_{xy})p(x \mid y, w_x, w_y, w_{xy}).$$

The first equality is due to the fact that x and w_y are conditionally independent given (w_x, w_y, w_{xy}), and y and w_x are conditionally independent given (w_y, w_x, w_{xy}). (See Figure 9.3.5.) The other two equalities follow from the product law.

Replacing $p(x \mid w_x, w_{xy})p(y \mid x, w_y, w_{xy})$ in the product of the conditional probability functions for the original diagram by

$$p(y \mid w_x, w_y, w_{xy})p(x \mid y, w_x, w_y, w_{xy})$$

we obtain the product of the conditional probability functions for the second diagram. This proves that the joint probability functions of the two diagrams are equal. Finally, we notice that if there were another directed path from x to y, we would create a cycle by reversing arc $[x, y]$, which is not allowed. □

In general, reversing an arc corresponds to applying Bayes' formula and the theorem of total probability. However, it may also involve the addition of arcs, and such arcs in some cases represent only pseudo dependencies. In this sense, some relevant information may have been lost after arc reversal.

The following example of the influence diagram for a fault tree OR gate illustrates the arc reversal operation. The deterministic node δ becomes probabilistic after arc $[x_1, \delta]$ is reversed.

Influence diagram for fault tree OR gate.

Let

$$\delta(x_1, x_2) = \begin{cases} 0 & \text{if } x_1 = x_2 = 0, \\ 1 & \text{otherwise} \end{cases}$$

$P(x_1 = 1) = p_1$ and $P(x_2 = 1) = p_2$. The influence diagram representation for the OR gate fault tree is

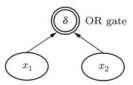

Reversing arc $[x_1, \delta]$ we have

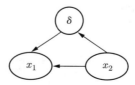

now with δ node conditional probabilities

	$P(\delta = 1 \mid x_2)$
$x_2 = 0$	p_1
$x_2 = 1$	1

and for $P(x_1 = 1 \mid \delta, x_2)$, we have the table

	$\delta = 0$	$\delta = 1$
$x_2 = 0$	0	1
$x_2 = 1$	undefined	p_1

Reversing the arc $[x_2, \delta]$, we have

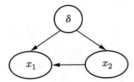

with δ node probability

$$P(\delta = 1) = p_1 + p_2 - p_1 p_2$$

and x_2 conditional probability given by the table

	$P(x_2 = 1 \mid \delta)$
$\delta = 0$	0
$\delta = 1$	$\dfrac{p_2}{p_1 + p_2 - p_1 p_2}$

Exercises

9.3.1. Fill in the conditional probabilities in the following influence diagram for a fault tree **AND** gate. The fill-in should be similar to the **OR** gate illustration.

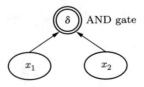

$P(x_1 = 1) = p_1; P(x_2 = 1) = p_2; x_1$ and x_2 are independent

$$\delta(x_1, x_2) = \begin{cases} 1 & \text{if } x_1 = x_2, \\ 0 & \text{otherwise.} \end{cases}$$

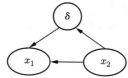

Fill in the table for $P(\delta = 1|x_2)$

	$P(\delta = 1|x_2)$
$x_2 = 0$	
$x_2 = 1$	

For $P(x_1 = 1 \mid \delta, x_2)$ fill in the table

	$\delta = 0$	$\delta = 1$
$x_2 = 0$		
$x_2 = 1$		

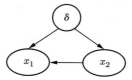

Calculate $P(\delta = 1)$ and fill in the table for $P(x_2 = 1 \mid \delta)$

	$P(x_2 = 1 \mid \delta)$
$\delta = 0$	
$\delta = 1$	

9.3.2. Are x_2 and x_3 conditionally independent of x_3 given x_1 in the following influence diagram?

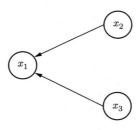

9.3.3. Suppose a lot of size N is taken from a production line. Assume in our judgment that

(1) each item produced in this line has the same chance π of being defective, and

(2) for any given value of π the items are independent relative to being defective.

Let x be the number found defective in a sample of size n and y the number defective in the remainder. What are the initial distributions of $p(x \mid \pi)$ and $p(y \mid \pi)$?

Reverse the relevant arcs in the right sequence in the following diagram to obtain $p(\pi \mid x)$ and $p(y \mid x)$. What are the distributions of $p(\pi \mid x)$ and $p(y \mid x)$?

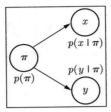

9.3.4. Unique lot example. Suppose a unique lot contains N items and that is all there are. Let θ be the number defective in the lot. Suppose a sample of size n is taken and x are found to be defective. Initially we have the influence diagram

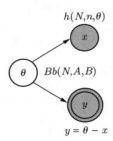

where $y = \theta - x$ is the number defective in the remainder and θ initially has a $Bb[N, A, B]$ prior distribution. By a sequence of arc reversals calculate the distribution of $p(y \mid x)$. (*Hint.* Review the beta-binomial distribution in section 3.1 of Chapter 3.)

9.3.5. Forensic science. A robbery has been committed and a suspect, a young man, is on trial. In the course of the robbery, a window pane was broken. The robber had apparently cut himself and a blood stain was left at the scene of the crime. Let x represent the blood type of the suspect, y the blood type of the blood stain found at the scene of the crime, and θ the quantity of interest, "the state of culpability" (guilt or innocence) of the suspect. Formally, and before using the actual values of the observable quantities, let

$$x = \begin{cases} 1 & \text{if the suspect's blood type is A,} \\ 0 & \text{otherwise;} \end{cases} \qquad y = \begin{cases} 1 & \text{if the blood stain} \\ & \text{type is A,} \\ 0 & \text{otherwise;} \end{cases}$$

$$\theta = \begin{cases} 1 & \text{if the suspect is guilty,} \\ 0 & \text{otherwise.} \end{cases}$$

The following diagram is an influence diagram constructed for this case. Note that the actual values of x and y that are known at the time of the analysis are not yet used. In fact, the diagram describes the dependence relations among these quantities and the conditional probabilities to be used.

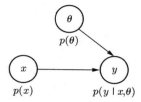

Influence diagram for a problem in forensic science.

Let π represent the proportion of people in the population with blood type A, and let ρ represent any jury member's probability that the suspect is guilty before the juror has learned about the blood stain evidence. Use the following probability assignments:

$$p(\theta) = \begin{cases} \rho & \text{if } \theta = 1, \\ 1 - \rho & \text{if } \theta = 0; \end{cases} \qquad p(x) = \begin{cases} \pi & \text{if } x = 1, \\ 1 - \pi & \text{if } x = 0; \end{cases}$$

$$p(y \mid x, y) = \begin{cases} \pi & \text{if } \theta \neq y = 1, \\ 1 - \pi & \text{if } \theta = y = 0, \\ 1 & \text{if } \theta = 1 \text{ and } y = x, \\ 0 & \text{otherwise.} \end{cases}$$

Using the influence diagram above and appropriate arc reversals, calculate the probability of guilt given the evidence, namely, $P(\theta = 1 \mid x = y = 1)$.

9.4*.   Conditional independence. The objective of this section is to study the concept of conditional independence and introduce its basic properties. We believe that the simplest and most intuitive way that this study can be performed is by using the total visual force of the probabilistic influence diagrams.

We now introduce the two most common definitions of conditional independence.

DEFINITION 9.4.1 (intuitive). *Given random quantities x, y, and z, we say that y is conditionally independent of x given z if the conditional distribution of y given (x, z) is equal to the conditional distribution of y given z.*

The interpretation of this concept is that, if z is given, no additional information about y can be extracted from x. The influence diagram representing this statement is presented in Figure 9.4.1.

DEFINITION 9.4.2 (symmetric). *Given random quantities x, y, and z, we say that x and y are conditionally independent given z if the conditional distribution of (x, y) given z is the product of the conditional distributions of x given z and that of y given z.*

The interpretation is that, if z is given, x and y share no additional information. The influence diagram representing this statement is displayed in Figure 9.4.2.

FIG. 9.4.1. *Intuitive definition of conditional independence.*

FIG. 9.4.2. *Symmetric definition of conditional independence.*

Using the arc reversal operation, we can easily prove that the probabilistic influence diagrams in Figures 9.4.1 and 9.4.2 are equivalent. Thus, Definitions 9.4.1 and 9.4.2 are equivalent, which means that in a specific problem we can use either one. To represent the conditional independence[2] described by both Figures 9.4.1 and 9.4.2, we can write either $x \perp y \mid z$ or $y \perp x \mid z$. This is a very general notation since x, y, and z are general random quantities (scalars, vectors, events, etc.). If in place of $\perp$ we use $\not\perp$, then x and y are said to be strictly dependent given z. We obtain independence (dependence) and write $x \perp y$ ($x \not\perp y$) if z is an event that occurs with probability one. It is important to notice that the symbol $\perp$ corresponds to the absence of an arc in a probabilistic influence diagram. However, the existence of an arc only indicates possible dependence. Although $\not\perp$ is the negation of $\perp$, the "absence of an arc" is included in the "presence of an arc."

The following proposition introduces the essence of the DROP/ADD principles for conditional independence, which are briefly discussed in the sequel.

PROPOSITION 9.4.3. *If $x \perp y \mid z$, then for every $f = f(x)$ we have*

(i) $f \perp y \mid z$ *and*

(ii) $x \perp y \mid (z, f)$.

The proof of this property is the sequence of diagrams of Figure 9.4.3. First note that (by the splitting nodes operation) to obtain the second diagram from the first we can connect to x a deterministic node f using arc $[x, f]$ without changing the joint probability function. Consequently, by reversing arc $[x, f]$ we obtain the third diagram. To obtain the last diagram from the third we use the merging nodes operation. Relations (i) and (ii) of Proposition 9.4.3 are represented by the second and the third diagrams of Figure 9.4.3.

As direct consequences of Proposition 9.4.3 we have

(C1) If $g = g(z)$ then $x \perp y \mid z$ if and only if $x \perp (y, g) \mid z$.

(C2) Let $f = f(x, z)$ and $g = g(y, z)$. If $x \perp y \mid z$ then $f \perp g \mid z$ and $x \perp y \mid (z, f, g)$.

---

[2]In the literature, the symbol II is often used instead of $\perp$.

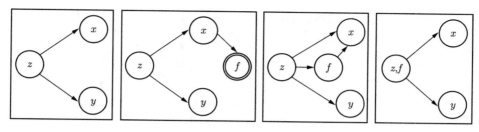

FIG. 9.4.3. *Proof of Proposition 9.4.3.*

The concept of conditional independence gives rise to many questions. Among them are the ones involving the DROP/ADD principles that we describe next. Suppose that x, y, z, w, f, and g are random objects such that $x \perp y \mid z$, $f = f(x)$ and $g = g(z)$. What can be said about the relation $\perp$ if f is substituted for x, g substituted for z, (y, w) substituted for y, or (z, w) substituted for z? In other words, can x, y, and z be reduced or enlarged without destroying the $\perp$ relation? In general, the answer is no. However, for special kinds of reductions or enlargements the conditional independence relation is preserved.

First, we present two simple examples to show that arbitrary enlargements of x, y, or z may destroy the $\perp$ relation. Consider now that w_1 and w_2 are two independent standard normal random variables; i.e., $w_1 \sim w_2 \sim N(0, 1)$, and $w_1 \perp w_2$. If $x = w_1 - w_2$ and $y = w_1 + w_2$, then $x \perp y$ but certainly $x \not\perp y \mid w_2$. Note that if z is a constant and $w = w_1$, we conclude that $x \perp y \mid z$ but $x \not\perp y \mid (z, w)$.

Second, we present an example to show that an arbitrary reduction of z, the conditioning quantity, can destroy the $\perp$ relation. Let w_1, w_2, and w be three mutually independent standard normal random quantities; i.e., $w_1 \perp (w_2, w)$, $(w_1, w_2) \perp w$, $w_2 \perp (w_1, w)$, $w_1 \perp w_2$, $w_1 \perp w$, $w_2 \perp w$, and $w_1 \sim w_2 \sim w \sim N(0, 1)$. Define $x = w_1 - w_2 + w$, $y = w_1 + w_2 + w$, and note that $x \perp y \mid w$ but $x \not\perp y$. As before, if z is a constant we can conclude that $x \perp y \mid (z, w)$ but $x \not\perp y \mid z$.

The destruction of the $\perp$ relation by reducing or enlarging its arguments is known as Simpson's paradox (for more details, see Lindley and Novick (1981)). The paradox, however, is much stronger since highly positively correlated random variables could be highly negatively correlated after some DROP/ADD operations. For instance, let z and w be two independent normal random variables with zero means. Define $x = z + w$, $y = z - w$ and note that the correlation between x and y is given by correlation $(x, y) = (1 - r)(1 + r) - 1$, where r is equal to the variance of w divided by the variance of z. Also, if z is given it is clear that the conditional correlation is -1. In order to make $COV(x, y)$ close to 1, we can consider r arbitrarily small. This shows that we can have cases where x and y are strongly positive (negative) dependent but, when z is given, x and y turn to be strongly negative (positive) conditionally dependent.

The following is another important property of conditional independence. It is presented in Dawid (1979).

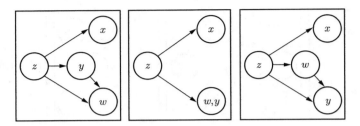

FIG. 9.4.4.　*Proof of Proposition 9.4.4.*

PROPOSITION 9.4.4. *The following statements are equivalent:*
  (i) $x \perp y \mid z$ *and* $x \perp w \mid (y, z)$,
 (ii) $x \perp (w, y) \mid z$, *and*
(iii) $x \perp w \mid z$ *and* $x \perp y \mid (w, z)$.

Figure 9.4.4 is the proof of Proposition 9.4.4. Again, only the basic probabilistic influence diagrams operations are used. The second graph is obtained from the first by merging nodes w and y. The third graph is obtained from the second by splitting node (w, y), and the first is obtained from the third by reversing arc $[w, y]$.

The above simple properties are very useful in some statistical applications, and they are related to the concept of sufficient statistic.

9.5.　Notes and references.

Section 9.1. Influence diagrams with decision nodes were invented in 1976 by Miller et al. (cf. Howard and Matheson (1984)).

Section 9.2. The discussion in this chapter was developed together with Carlos Pereira and appeared in Barlow and Pereira (1991).

Section 9.3. The arc reversal operation was extensively investigated by Shachter (1986). There have been a number of computer programs developed to analyze influence diagrams including decision nodes. One of the earliest was developed by Shachter and called DAVID. A decent free version and more or less a superset of DAVID's functionality is Netica. Netica runs on PCs and Macintoshes. Microsoft has a fairly nice system with a educational free license, but only for PCs, called MSBN. See http://www.auai.org/.

Section 9.4. This discussion is related to the paper by Basu and Pereira (1982).

Notation

$x \perp y$　　　x and y are independent
$x \perp y \mid z$　　x and y are independent conditional on z
$x \not\perp y$　　　x and y are dependent
$x \not\perp y \mid z$　　x and y are dependent given z

CHAPTER **10**

Making Decisions Using Influence Diagrams

10.1. Introduction.

The previous chapter was restricted to questions that could be answered by computing conditional probabilities, that is, the calculation of probabilities on the receipt of data, or by information processing. This chapter is concerned with the evaluation and use of information through the use of influence diagrams with decision nodes and value nodes. In section 10.5 we compare influence diagrams and decision trees. Both can be used for decision making.

In real life we are continually required to make decisions. Often these decisions are made in the face of a great deal of uncertainty. However, time and resources (usually financial) are the forcing functions for decision. That is, decisions must be made even though there may be a great deal of uncertainty regarding the unknown quantities related to our decision problem.

In considering a decision problem, we must first consider those things that are known as well as those things that are unknown but are relevant to our decision problem. It is very important to restrict our analysis to those things that are relevant, since we cannot possibly make use of all that we know in considering a decision problem. So, the first step in formulating a decision problem is to limit the universe of discourse for the problem.

A decision problem begins with a list of the possible alternative decisions that may be taken. We must seriously consider all the exclusive decision alternatives that are allowed. That is, the set of decisions should be exhaustive as well as exclusive. We then attempt to list the advantages and disadvantages of taking the various decisions. This requires consideration of the possible uncertain events related to the decision alternatives. From these considerations we determine the consequences corresponding to decisions and possible events. At this point, in most instances, the decisions are "weighed" and that decision is taken that is deemed to have the most "weight." It is this process of "weighing" alternative decisions that concerns us in this chapter.

An important distinction needs to be made between decision and outcome. A good outcome is a future state of the world that we prefer relative to other possibilities. A good decision is an action we take that is logically consistent with the alternatives we perceive, the information we have, and the preferences we feel at the time of decision. In an uncertain world, good decisions could lead to bad outcomes and bad decisions can conceivably lead to good outcomes. Making the distinction allows us to separate action from consequence. The same distinction needs to be made between prior and posterior. In retrospect, a prior distribution may appear to have been very bad. However, based on prior knowledge alone, it may be the most logical assessment. The statement, "Suppose you have a 'bad' prior" is essentially meaningless unless "bad" means that a careful judgment was not used in prior assessment.

The purpose of this chapter is to introduce a "rational method" for making decisions. By a "rational method," we mean a method which is internally consistent—that is, it could never lead to a logical contradiction. The method we will use for making decisions can be described in terms of influence diagrams. So far we have only discussed probabilistic influence diagrams.

10.2. Decision nodes.

Probabilistic influence diagrams needed only probabilistic nodes, deterministic nodes, and directed arcs. For decision making we also need decision nodes and value nodes. The following example demonstrates the need for these additional nodes.

Example 10.2.1 (two-headed coins). Suppose your friend tells you that he has a coin that is either a "fair" coin or a coin with two heads. He will toss the coin and you will see which side comes face up. If you correctly decide which kind of a coin it is, he will give you one dollar. Otherwise you will give him one dollar. If you accept his offer, what decision rule should you choose? That is, based on the outcome of the toss, what should your decision be? In terms of probabilistic influence diagrams we only have Figure 10.2.1,

where

$$\theta = \begin{cases} \text{fair if the coin is fair,} \\ \text{two-headed otherwise;} \end{cases}$$

and

$$x = \begin{cases} T & \text{if the toss results in a tail,} \\ H & \text{otherwise.} \end{cases}$$

Fig. 10.2.1.

Clearly, if $x = T$, you know the coin does not have two heads. The question is, what should you decide if $x = H$? To solve your problem, we introduce a decision node which is represented by a box or rectangle. The decision node represents the set of decisions that can be taken; namely,

$$d_1: \quad \text{decide the coin is fair,}$$
$$d_2: \quad \text{decide two-headed.}$$

Attached to the decision node is a set of allowed *decision rules*, which depend on the outcome of the toss x. For example, one decision rule might be

$$\delta(x) = \begin{cases} d_1 & \text{if } x = T, \\ d_2 & \text{if } x = H. \end{cases}$$

Another decision rule might be

$$\delta(x) = d_1 \text{ for all } x.$$

The influence diagram helpful for solving our problem is Figure 10.2.2.

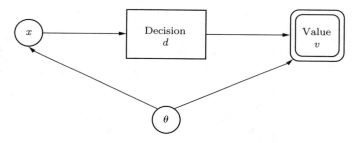

FIG. 10.2.2. *Two-headed coins.*

In this diagram, the double bordered node represents a value or utility node. The value node represents the set of consequences corresponding to possible pairs of decisions and states (d, θ). Attached to the value node is the value function $v(d, \theta)$, a deterministic function of the states of adjacent predecessor nodes. In this example,

$$v(d, \theta) = \begin{cases} \$1 & \text{if } d = d_1 \text{ and } \theta = \text{fair or } d = d_2 \text{ and } \theta = 2 \text{ heads,} \\ -\$1 & \text{otherwise.} \end{cases}$$

The reason for initially drawing the arc $[\theta, x]$ rather than the arc $[x, \theta]$ is that, in general, it is easier to first assess $p(\theta)$ and $p(x \mid \theta)$ rather than to directly assess $p(x)$ and $p(\theta \mid x)$.

The optimal decision will depend on our initial information concerning θ, namely, $p(\theta)$. However, since θ is unknown at the time of decision, there is no arc $[\theta, d]$. At the time of decision, we know x but not θ. Input arcs to a decision node indicate the information available at the time of decision. In general, there can be more than one decision node, as the next example illustrates.

Example 10.2.2 (sequential decision making). Consider an urn containing white and black balls. Suppose we know that the proportion of white balls θ is either $\theta = \frac{1}{3}$ or $\theta = \frac{2}{3}$, but we do not know which. Our problem is to choose between two actions. One action, say, a_1, would be to say $\theta = \frac{2}{3}$, while a_2 would be to say $\theta = \frac{1}{3}$. If we are wrong, we lose one dollar. Otherwise, we lose nothing. We can, if we choose, first draw a ball from the urn at cost c so as to learn more about θ. After observing the color of the ball drawn, say, x, then we must choose either action a_1 or a_2 at cost $(1 + c)$ if we are wrong and only cost c if we are right.

The first decision can be either d_1: take action a_1, d_2: take action a_2, or d_3: draw a ball from the urn. If we draw a ball from the urn, then our second decision after the drawing depends on the ball drawn x and must be either (1) taken action a_1 or (2) take action a_2. In this problem there are two decision points, and a second decision is needed only if the first decision is to continue sampling. Figure 10.2.3 is the decision influence diagram for our problem.

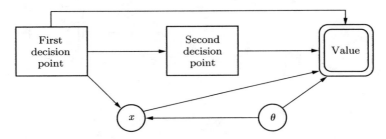

FIG. 10.2.3. *Sequential decision making.*

Example 10.2.3 (inspection sampling). The following example is a decision problem of some practical importance and is based in part on Deming's inspection problem in Chapter 5, section 5.3. Periodically, lots of size N of similar units arrive and are put into assemblies in a production line. The decision problem is whether or not to inspect units before they are put into assemblies. If we opt for inspection, what sample size n of the lot size N should be inspected? In any event, haphazard sampling to check on the proportion defective in particular lots is prudent.

Let π be the percent defective over many lots obtained from the same vendor. Suppose we believe that the vendor's production of units is in statistical control. That is, each unit, in our judgment, has the same chance π of being defective or good regardless of how many units we examine. Let $p(\pi)$ be our probability assessment for the parameter π based on previous experience. It could, for example, be degenerate at, say, π_0.

Let k_1 be the cost to inspect one unit before installation. Let k_2 be the cost of a defective unit that gets into the production line. This cost will include the cost to tear down the assembly at the point where the defect is discovered. If a unit is inspected and found defective, additional units from another lot are inspected until a good unit is found. (We make this model assumption since all defective units that are found will be replaced at vendor's expense.) Figure 10.2.4 illustrates our production line problem.

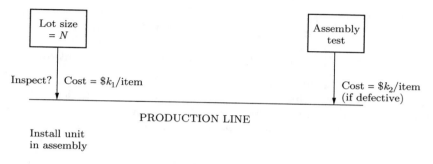

FIG. 10.2.4. *Deming's inspection problem.*

We assume the inspection process is perfect; i.e., if a unit is inspected and found good, then it will also be found good in the assembly test.

The all or none rule. It has been suggested [cf. Deming (1986)] that the optimal decision rule is always to either choose $n = 0$ or $n = N$, depending on the costs involved and π, the chance that a unit is defective. Later we will show that this is not always valid.

   In this example we consider the problem under the restriction that the initial inspection sample size is either $n = 0$ or $n = N$. The decisions are $n = 0$ and $n = N$. Figure 10.2.5 is an influence diagram for this problem, where

   x = number of defectives in the lot and

   y = number of additional inspections required to find good
        replacement units for bad units.

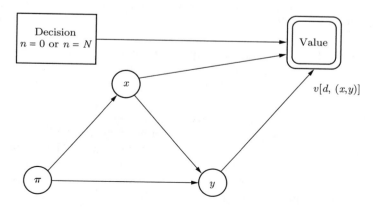

FIG. 10.2.5. *Influence diagram for Deming's inspection problem.*

The value (loss) function is

$$v[d,(x,y)] = \begin{cases} k_1 y + k_2 x & \text{if } n = 0, \\ k_1 N + k_1 y & \text{if } n = N. \end{cases}$$

We have already solved this problem in Chapter 5, section 5.3. The solution is that $n = 0$ is best if $E[\pi] < \frac{k_1}{k_2}$ and $n = N$ otherwise. See Figure 10.2.5.

10.3.   Formal definitions and influence diagram construction.

Definitions and basic results.   A decision influence diagram (or influence diagram for short) is a diagram helpful in solving decision problems. We have introduced two new types of nodes, namely, decision nodes and value nodes. Figure 10.3.1 is a typical decision node with input and output arcs.

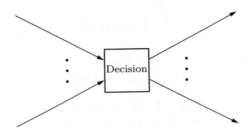

FIG. 10.3.1.   *Decision node.*

DEFINITION 10.3.1.   *A decision node*
(1) *represents the possible decisions that may be taken at a given point in time,*
(2) *and attached to each decision node is a set of allowed decision rules or mappings from possible states of adjacent predecessor nodes to the set of possible alternative decisions represented by the node itself.*

For example, there could be only one *allowed decision rule* corresponding to a given decision node. In this case the decision node can be replaced by a deterministic node.

DEFINITION 10.3.2.   *A decision rule δ corresponding to a decision node is a mapping from possible states of adjacent predecessor nodes (deterministic and probabilistic as well as previous decisions) to a set of possible alternative decisions.*

Decision nodes will, in general, have both directed input and directed output arcs. What makes decision nodes very different from probabilistic and deterministic nodes is that these arcs may never by reversed. Adjacent predecessor nodes to a decision node indicate information available to the decision maker at the time of that particular decision. The value of an adjacent predecessor node to a decision node is known at the time of decision; it represents certain

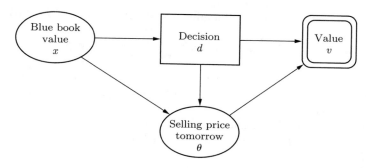

FIG. 10.3.2.

knowledge, not possible knowledge. In this sense the arc is different from arcs between probability nodes that can indicate only possible dependence. Since decision nodes imply a time ordering, the corresponding directed arcs can never be reversed.

Directed arcs emanating from a decision node denote possible dependence of adjacent successor nodes on the decision taken. These arcs can likewise never be reversed.

Example 10.3.3 (selling a car). Suppose you plan to sell your car tomorrow, but the finish on your car is bad. Your decision problem is whether or not to have your car painted today in order to increase the value of your car tomorrow. Note that tomorrow's selling price depends on today's decision d. Let x be yesterday's blue book value for your car and c the cost of a paint job. Let θ be the price you will be able to obtain for your car tomorrow. Let $v(d, \theta)$ be the value to you of today's decision d and tomorrow's selling price θ. The influence diagram associated with your decision problem is shown in Figure 10.3.2. Obviously you cannot reverse arc $[x, d]$, since today's decision cannot alter yesterday's blue book value. Likewise, you cannot reverse arc $[d, \theta]$, since we cannot know today, for sure, what tomorrow will bring.

The value node is similar to a deterministic node. What makes it different is that it has no successor nodes.

DEFINITION 10.3.4. *A value node is a sink node that*

(1) *represents possible consequences corresponding to the states of adjacent predecessor nodes (a consequence can, for example, be a monetary loss or gain),*

(2) *has an attached utility or loss function which is a deterministic function consequence represented by the value node itself.*

A value node shares with a decision node the property that input arcs cannot be reversed. However, if a value node v has a probabilistic adjacent predecessor node, say, θ, and v is the only adjacent successor of θ (as in Figure 10.3.2), then node θ can be eliminated.

We now give a formal definition of a decision influence diagram.

DEFINITION 10.3.5. *A decision influence diagram is an acyclic directed graph in which*

(i) *nodes represent random quantities, deterministic functions, and decisions;*

(ii) *directed arcs into probabilistic and deterministic nodes indicate possible dependence, while directed arcs into decision nodes indicate information available at the time of decision;*

(iii) *attached to each probabilistic (deterministic) node is a conditional probability (deterministic) function, while attached to each decision node is a set of allowed decision rules;*

(iv) *decision nodes are totally ordered and there is a directed arc (perhaps implicitly) to each decision node from all predecessor decision nodes;*

(v) *there is exactly one deterministic sink node called the value node.*

Using decision influence diagrams to model problems. In Example 10.2.1 (two-headed coins) you were offered the opportunity to make a dollar also with the risk of losing a dollar. Your allowable decisions were

$$d_1: \quad \text{decide the coin is fair,}$$

$$d_2: \quad \text{decide two-headed.}$$

Hence you might start by drawing the decision node d with the two allowable choices d_1 and d_2. You might then draw the value node v, since the value node denotes the objective of your decision analysis. Your objective is to calculate $v(d)$, the unconditional value function, as a deterministic function of the decision taken d. However, it may be easier to first determine the conditional value function as a deterministic function of relevant unknown quantities as well as perhaps prior decisions relevant to your problem.

DEFINITION 10.3.6. *A value function is called unconditional if it only depends on the decision taken and not on relevant unknown quantities. It is called conditional when it also depends on relevant unknown quantities.*

Since it is easier to think of the value function as a conditional deterministic function of your decision and the property of the coin, say, θ, you also need to draw a node for θ. Since θ is unknown to you, node θ is a probabilistic node. Since the value node v depends on both the decision taken d as well as the property of the coin θ, draw arcs $[d, v]$ and $[\theta, v]$. Attach a deterministic function $v(d, \theta)$ to node v. Figure 10.3.3 shows the influence diagram at this stage of the analysis.

You are allowed to see the result of one coin toss, and this information is available at the time you make your decision. Hence draw a node x for the outcome of the toss. Since the outcome of the toss (before you see it) is an unknown random quantity for you, draw a probabilistic node for x. Since the probability function for x depends on θ, draw arc $[\theta, x]$ and assess $p(x \mid \theta)$ and $p(\theta)$. Draw arc $[x, \delta]$ since you will know x at the time you make your decision d. The diagram now looks like Figure 10.2.2, which we repeat below.

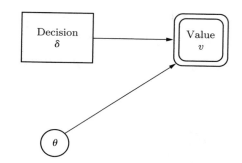

FIG. 10.3.3. *First stage of diagram development.*

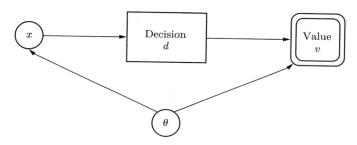

FIG. 10.2.2 *(repeated).*

10.4.   Decision influence diagram operations.

There are essentially two influence diagram graph operations that are used to solve decision problems expressed as influence diagrams. They are
(1) the elimination of probabilistic nodes and
(2) the elimination of decision nodes.
In the process of eliminating probability nodes you will obtain the unconditional value function $v(d)$ as a function of decisions allowed. Having obtained $v(d)$ you can determine the optimal decision with respect to the unconditional value function and in this sense eliminate the decision node. The justification for both graph operations is based on the idea that you should be self-consistent in making decisions.

Solution of the two-headed coin problem.   Having described how to construct the influence diagram (Figure 10.2.2, for the two-headed coin Example 10.2.1) we now discuss its solution. Our objective is to calculate $v[\delta(x)]$, where $\delta(x)$ depends on x and is either

$$d_1:\quad \text{decide the coin is fair}$$

or

$$d_2:\quad \text{decide two-headed.}$$

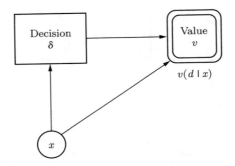

FIG. 10.4.1. *Elimination of the decision node.*

You must eliminate θ since, as it stands, the optimal decision would depend on θ, which is unknown. If the arc $[\theta, v]$ were the only arc emanating from θ you would be able to do this immediately by summation or by integration. Since there is also another arc, namely, $[\theta, x]$, emanating out of θ, this is not possible.

Elimination of node θ.   Since nodes x and θ are probabilistic nodes, you can reverse arc $[\theta, x]$ using the arc reversal operation. If you do this, $p(x \mid \theta)$ attached to node x is changed to $p(x)$ using the theorem of total probability while $p(\theta)$ is changed to $p(\theta \mid x)$ using Bayes' formula. After this arc reversal, arc $[\theta, v]$ is the only arc emanating from node θ and node θ can now be eliminated by summing $v[\delta, \theta]$ with respect to $p(\theta \mid x)$. This results in the influence diagram of Figure 10.4.1. The deterministic function attached to node v is now

$$v[\delta(x)] = v[\delta(x), \theta = \text{fair}]p(\theta = \text{fair} \mid x)$$
$$+v[\delta(x), \theta = 2 \text{ headed}]p(\theta = 2 \text{ headed} \mid x).$$

See Figure 10.4.1.

To solve your problem you need only eliminate the decision node. This is accomplished by maximizing $v[\delta(x)]$ over decision alternatives since you want to maximize your winnings.

Example. Inspection sampling.   In Chapter 5, section 5.3 and in Example 10.2.3 in this chapter we discussed Deming's inspection problem under the restriction that the sample size n was either $n = 0$ or $n = N$, the lot size. Figure 10.4.2 is the decision influence diagram for this problem when the sample size can be any integer between 0 and N.

We leave the solution of this example to Exercise 10.4.2.

Example. Value added to substrate.   Deming has a variation of the inspection sampling problem called "value added to substrate." Work is done on incoming material, the substrate. The finished product will be classified as first grade, second grade, third grade, or scrap. If the incoming item is not defective, then it will result in a final product of first grade. Otherwise, if it is not

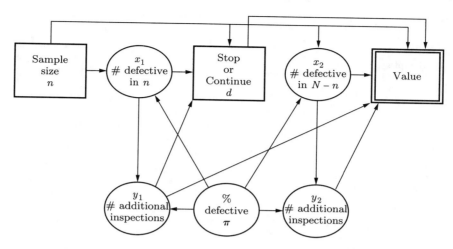

FIG. 10.4.2. *Influence diagram for Deming's inspection problem.*

inspected at the point of installation and if it is defective, then a final product of downgrade or scrap will be produced. Every final product is subject to assembly grading. The only difference between this model and the previous model is that no additional inspections occur at assembly test. This is because the item (say a bag of cement) cannot be replaced at assembly test if it is not inspected at the point of installation but is found defective at the point of assembly test. As before, let k_1 be the cost of inspecting an item. Let k_2 be the average loss from downgrading the final product or for scrapping finished items.

Suppose that π, the percent defective over many lots, has a prior $Be(A, B)$ distribution. We will first modify the decision influence diagram in Figure 10.4.2 and then, without numerically solving the problem, show the sequence of arc reversals and decision making required to calculate the optimal inspection sample size n.

Solution of value added to substrate. Delete node y_2 and all arcs into and out of this node in Deming's first model. Note that when $n = N$, the STOP or CONTINUE decision node does not apply since there is no remainder to inspect. Also when $n = N$, $x_2 \equiv 0$ since there can be no defectives in an empty sample.

In general, we *should not* have an arc from a decision node to a probability node. In the case of the x_2 node, the distribution of x_2 does not depend on the decision taken but does depend on n and π. The technical name of this requirement is the "sure thing principle" or axiom 0 in Appendix B. Node n is a special case since it is the initial condition determining sampling distributions. See Figure 10.4.3.

Algorithm for solving influence diagram Figure 10.4.3. A list ordering of the nodes is $n < \pi < x_1 < y_1 < d < x_2 < Loss$. Both n and π are root nodes. The list ordering is *not* unique.

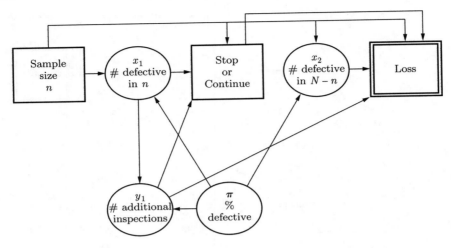

FIG. 10.4.3. *Influence diagram for value added to substrate.*

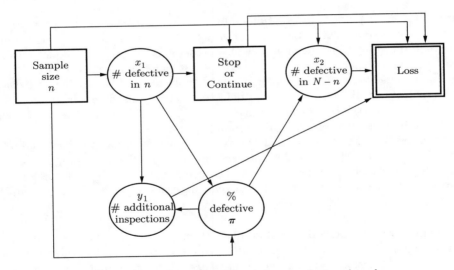

FIG. 10.4.4. *Influence diagram after reversal of arc* $[\pi, x_1]$.

First we eliminate π. Start by reversing arc $[\pi, x_1]$ and adding arc $[n, \pi]$. See Figure 10.4.4.

Now reverse arc $[\pi, y_1]$ and add arc $[n, y_1]$. Note that we could not have reversed $[\pi, y_1]$ before reversing arc $[\pi, x_1]$ since this would have created a cycle. See Figure 10.4.5.

Now we can eliminate node π by calculating the distribution of x_2 given n and y_1. If π is $Be(A, B)$, then π, given n and y_1, is $Be(A + y_1, B + n)$ while x_2, given y_1 and n, is $Bb(N - n, A + y_1, B + n)$. Note that x_2 does not depend on x_1. Actually π depends only on n (the total number of items inspected and found to be good) and y_1 (the total number of defective items found by inspection). See Figure 10.4.6.

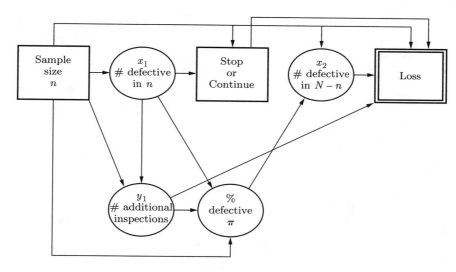

FIG. 10.4.5. *Influence diagram after reversal of arc* $[\pi, y_1]$.

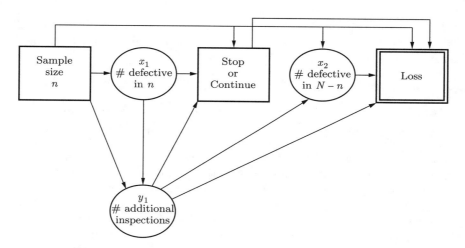

FIG. 10.4.6. *Influence diagram after eliminating node* π.

At this point we can calculate the expected loss if we STOP and also if we CONTINUE, namely,

if we STOP, the LOSS is $k_1(n + y_1) + k_2 E[x_2 \mid n, y_1]$;

if we CONTINUE, the LOSS is $k_1(N + y_1)$.

We take that decision that has minimum expected loss given n, y_1, thus eliminating the STOP or CONTINUE decision node. See Figure 10.4.7.

Next we calculate the minimum of the expected losses following either the STOP or CONTINUE policy:

$$\text{minimum}[k_1(n + y_1) + k_2 E[x_2 \mid n, y_1], k_1(N + y_1)].$$

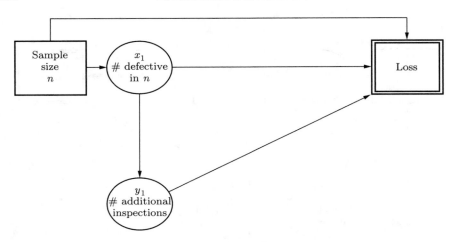

FIG. 10.4.7. *Influence diagram after eliminating a decision node.*

The minimum value determines the optimal STOP or CONTINUE policy. We calculate the expected value of this minimum with respect to the distribution of y_1 given x_1 and n.

Finally, we calculate the expected loss with respect to x_1, thus eliminating node x_1, and then calculate that value of n that minimizes the total expected loss.

Exercises

10.4.1. *Two-headed coins.* Let $p(\theta = fair) = \pi$ and determine your optimal decision rule as a function of the outcome of the coin toss x and π.

10.4.2. *Deming's inspection problem.* Figure 10.4.2 models Deming's inspection problem. Fix on n and $N > n$. Assume your prior for the percent defective π is $Be(A, B)$. The optimal solution should depend on the costs, $N, n, x_1,$ and y_1. Note that at the time of decision, you do *not* know either x_2 or y_2.

(a) Why is $p(\pi \mid x_1, y_1) = p(\pi \mid y_1)$?

(b) Determine the optimal inspection decision after sampling; i.e., should we *STOP* inspection or *CONTINUE* and inspect all of the remaining items?

10.4.3. Find the optimal decisions for the sequential decision-making example, Figure 10.2.3, as a function of the following prior for θ:

$$p\left(\theta = \frac{1}{3}\right) = \pi.$$

10.5. Comparison of influence diagrams and decision trees.

Decision trees provide another graphical representation of decision problems. In this case, the graph construction is based on the temporal order in which decisions are made and unknown quantities are actually or potentially revealed. The method is equivalent to enumerating all possible sample space outcomes;

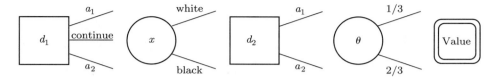

FIG. 10.5.1.   *Temporal order of a sequential decision tree.*

i.e., it is equivalent to sample space enumeration. The decision tree provides a more concrete representation of the problem than the influence diagram. In this respect the beginner may understand it more easily and usually prefers it to the influence diagram representation.

We will illustrate the decision tree methodology using the previous Example 10.2.2 concerning sequential analysis discussed in section 10.2.

Example 10.5.1. Sequential decision making.   Consider an urn containing white and black balls. Suppose we know that the proportion of white balls θ is either $\theta = \frac{1}{3}$ or $\theta = \frac{2}{3}$, but we do not know which. Our problem is to choose between two actions. One action, say, a_1, would be appropriate were $\theta = \frac{2}{3}$, while a_2 would be appropriate were $\theta = \frac{1}{3}$. If we are wrong, we lose one dollar. Otherwise, we lose nothing. We can, if we choose, first draw a ball from the urn at cost \$$c$ so as to learn more about θ. After observing the color of the ball drawn, say, x, then we must choose either action a_1 or a_2 at cost \$$(1 + c)$ if we are wrong and only cost \$$c$ if we are right.

In this problem there are two decision points, and a second decision is needed only if the first decision is to continue sampling. Figure 10.5.1 presents the decision process in temporal order beginning with the initial decision to stop and decide or to continue. If you decide to stop, then of course the color x of the ball is not observed and decision d_2 is not considered.

A full decision tree representation for the sequential analysis example is presented in Figure 10.5.2. A decision tree, drawn in temporal order as shown in Figure 10.5.2 is in effect a connected directed graph, where each arc is directed from left to right. Considered as a directed tree it has exactly one root. In Figure 10.5.2 the root is the initial decision node on the left. The terminal nodes $(L_1, L_2, \ldots, L_{12})$ are called the leaves of the directed graph. Unlike the influence diagram, the arcs in the decision tree carry the most information. Rectangles denote decision nodes while circles denote random nodes. Random node θ in Figure 10.5.1 is actually represented by six circles on the right in Figure 10.5.2. The leaves in Figure 10.5.2 correspond to possible values of the value node in Figure 10.5.1.

In order to give a formal definition of a decision tree, we first define a graph structure called a tree.

DEFINITION 10.5.1.   *An undirected connected graph with nodes or vertices V and (undirected) arcs joining the nodes is called a* tree *if it has no cycles; i.e., there is no sequence of connected nodes $i_1, i_2, \ldots, i_k$ such that nodes i_1 and i_k*

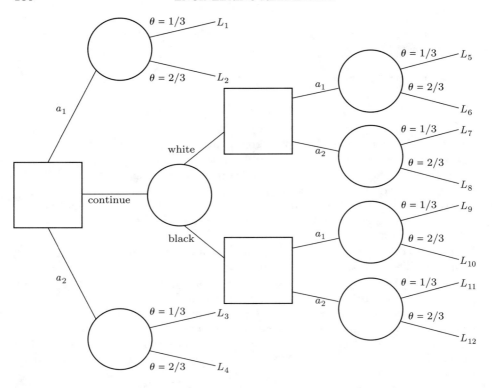

FIG. 10.5.2. *Decision tree for the sequential analysis example.*

are identical. A directed graph is a tree if the corresponding undirected graph is a tree.

Influence diagram graphs are not, in general, trees in the graph theory sense since the corresponding undirected graph may contain cycles in the sense of Definition 10.5.1.

A path to a node in a decision tree from the root node consists of a connected sequence of nodes and arcs from the root node to that node. The fundamental property of trees used in decision tree analysis is that there is a unique path from the root node to any other specified node. For example

$$\left\{ \text{decision} = \text{continue}, x = \text{white}, \text{decision} = a_1, \theta = \frac{1}{3} \right\}$$

is the unique path from the root node to the top right-hand leaf node in Figure 10.5.2. There are 12 such paths corresponding to the 12 leaves in Figure 10.5.2. All possible paths corresponding to leaves constitute the sample space. The decision tree diagram can be used to enumerate all such possible states in the sample space. Obviously, the paths to leaf nodes constitute a set of mutually exclusive and exhaustive states corresponding to our decision problem.

DEFINITION 10.5.2. *A decision tree is a rooted directed graph that is a tree in the graph theory sense and in which*

(i) *nodes represent decisions, random quantities, and consequences;*

(ii) *arcs emanating from a decision node correspond to the possible decisions which may be taken at that node (possible decisions corresponding to a decision node may depend on the unique path from the root node to that decision node);*

(iii) *arcs emanating from a random node correspond to possible realizations of the corresponding random quantity (possible realizations may depend on the unique path from the root node to that random node);*

(iv) *a leaf of the tree corresponds to a consequence resulting from the realization of arc events on the path from the root node to that leaf node.*

Comparison of decision trees and influence diagrams. The decision tree will have many more nodes and arcs than the corresponding influence diagram. It is, however, more helpful in visualizing all possible outcomes of the decision problem. In the influence diagram, Figure 10.2.3, a decision or random quantity is represented by one node. In the decision tree, Figure 10.5.2, it may be represented by many nodes. Also it is not as easy to recognize conditional independence in a decision tree as it is in the influence diagram. However, both representations can be used to solve decision problems, and the solution process is essentially the same.

Note that the decision tree shown in Figure 10.5.2 is *symmetric*. An influence diagram corresponds to a symmetric decision tree. There are decision problems for which using the corresponding decision tree may be more efficient since there may be computational savings to be achieved through asymmetric decision tree processing.

In general, we recommend the use of influence diagrams. Using decision trees it may be difficult to see the forest (the main influences in the decision problem) through the trees (the possible sample space path realizations).

10.6. Notes and references.

Section 10.1. An excellent discussion of decision influence diagrams is the paper, "From Influence to Relevance to Knowledge" by R. A. Howard (1990).

Section 10.2. The two-headed coin example is due to Carlos Pereira. Example 10.2.2 on sequential decision making is considered in much greater detail in de Finetti (1954). He considers the problem of combining different opinions in that paper.

Section 10.3. The intent of this section is to demonstrate the thinking process behind influence diagram construction.

Section 10.4. The intent of this section is to demonstrate the thinking process behind solving influence diagrams. The algorithmic process is called backward induction since we start with the value node. There are several computer programs available for solving decision influence diagrams. We mentioned Netica and MSBN in Chapter 9, section 9.5.

Section 10.5. An introduction to decision trees is also available in Lindley (1985).

Classical Statistics Is Logically Untenable

We have not mentioned in this book most of the methodology of so-called classical statistics. The reason is that classical statistics is based on deductive analysis (the logic of mathematics), whereas statistical inference and decision theory are concerned with inductive analysis (probability judgments). Why is classical statistics logically untenable? D. Basu (1988), in his careful examination of the methodology due to R. A. Fisher, summed up his answer to this question as follows: "It took me the greater part of the next two decades to realize that statistics deals with the mental process of induction and is therefore essentially antimathematics. How can there be a deductive theory of statistics?"

To be more concrete, we consider two favorite ideas of classical statistics, namely, unbiasedness and confidence intervals.

Unbiasedness. The posterior mean is a Bayes estimator of a parameter, say, θ, with respect to squared error loss. It is also a function of the data. No Bayes estimator (based on a corresponding proper prior) can be unbiased in the sense of classical statistics; see Bickel and Blackwell (1967).

An estimator $\hat{\theta}$ (a function of data) is called unbiased in the sense of classical statistics if

$$E_F[\hat{\theta} \mid \theta] = \theta$$

for each $\theta \in \Theta$. In this case, F is the probability distribution corresponding to sample data.

Most unbiased estimators are in fact inadmissible with respect to squared error loss in the sense of classical statistics. D. Basu provides the following example. In the case of the exponential density

$$\frac{1}{\theta} e^{-\frac{x}{\theta}},$$

$\hat{\theta} = \frac{T}{k}$ is an unbiased estimator for θ in the sense of classical statistics, where T is the total time on test (TTT) and k is the number of observed failures. However, it is inadmissible in the sense that there exists another estimator $c\hat{\theta}$ with $c \neq 1$

such that for all θ,

$$E_F[(c\hat{\theta} - \theta)^2 \mid \theta] < E_F[(\hat{\theta} - \theta)^2 \mid \theta].$$

To find this c, consider $Y = \frac{\hat{\theta}}{\theta}$, and note that $E(Y) = 1$. Then we need only find c such that

$$E_F[(cY - 1)^2 \mid \theta]$$

is minimum. This occurs for $c = \frac{E(Y)}{E(Y^2)}$, which is clearly not 1. Hence $\hat{\theta}$ is inadmissible in the sense of classical statistics.

Unbiasedness is not a viable criterion even for classical statisticians. Statisticians who require unbiasedness and admissibility are not self-consistent.

Confidence intervals.   A $(1-\alpha)100\%$ confidence interval in the sense of classical statistics is one such that if the experiment is repeated infinitely often (and the interval recomputed each time), then $(1-\alpha)100\%$ of the time the interval will cover the fixed unknown "true" parameter θ.

This is *not* what most engineers mean by the term "confidence interval." A posterior probability interval, conditional on the data, is what most engineers have in mind. A $(1-\alpha)100\%$ confidence interval is not a probability interval conditional on the data.

Since confidence intervals do not produce a probability distribution on the parameter space for θ, they cannot provide the basis for action in the decision theory sense; i.e., a decision maker cannot use a confidence interval in the sense of classical statistics to compute an expected utility function which can then be maximized over the decision maker's set of possible decisions.

Suppose we choose the improper prior $p(\lambda) = \frac{1}{\lambda}$ for the parameter λ of the exponential density $\lambda e^{-\lambda x}$. In this case, the chi-square smallest $(1-\alpha)100\%$ probability interval and the confidence interval, in the sense of classical statistics, agree. Unfortunately, such improper probability intervals can be shown to violate certain rules of logical behavior. Dennis Lindley provides the following simple illustration of this fact.

Suppose n items are put on life test and we stop at the first failure, so that the TTT is $T = nx_{(1)}$. Now T given λ also has density $\lambda e^{-\lambda x}$ so that $\frac{\ln 2}{T}$ is a 50% improper upper probability limit on λ; i.e.,

$$p(\lambda \mid T) \propto p(T \mid \lambda)p(\lambda) = e^{-\lambda T}$$

so that $p(\lambda \mid T) = Te^{-\lambda T}$ and

$$P\left[\lambda \leq \frac{\ln(2)}{T} \mid T\right] = 0.50.$$

Suppose we accept this probability statement and that T is now observed. Consider the following hypothetical bet:
  (i) if $\lambda < \frac{\ln 2}{T}$ we lose the amount e^{-T};
  (ii) if $\lambda \geq \frac{\ln 2}{T}$ we win the amount e^{-T}.

We can pretend that the "true" λ is somehow revealed and bets are paid off. If we believe the probability statement above, then such a bet is certainly fair, given T.

Now let us compute our expected gain before T is observed (preposterior analysis). This is easily seen to be (conditional on λ)

$$-\int_0^{\ln 2/\lambda} \lambda e^{-\lambda t} e^{-t} dt + \int_{\ln 2/\lambda}^{\infty} \lambda e^{-\lambda t} e^{-t} dt = \frac{\lambda}{1+\lambda}[2^{-\frac{1}{\lambda}} - 1],$$

which is negative for all $\lambda > 0$. Note that this is what we would expect were we to make this bet infinitely often.

But this situation is again not self-consistent. Classical statistics is not self-consistent though it does claim to be "objective" and "scientific." Objectivity is really nothing more than a consensus, and scientists use induction as well as deduction to extrapolate from experiments.

APPENDIX **B**

Bayesian Decision Analysis
Is Self-Consistent

The influence diagram operations developed in Section 10.4 depend on the principle that you should be self-consistent in making decisions. Bayesian decision analysis is, above all, based on the premise that you should, at the very least, be self-consistent in the way in which you make decisions. Self-consistency in decision making has been defined by H. Rubin (1987) in terms of what he calls the *axioms of rational behavior.*

Of course, by self-consistency we do not mean that you are not capable of revising your opinion and possibly changing your decision (if allowed) upon receipt of new information. The Bayesian approach to decision analysis, which we use in this book is self-consistent. But what is, perhaps, even more important is that self-consistency implies Bayesian behavior.

The two influence diagram graph operations that we now justify are

(1) the elimination of decision nodes and
(2) the elimination of probability nodes.

Eliminating decision nodes: Maximizing unconditional utility. Self-consistency in decision making has been defined by H. Rubin (1987) in terms of what he calls the weak axioms. Given a decision problem, i.e., a set of possible alternative decisions and relevant information associated with the decision problem, the weak axioms (or self-consistency) imply that there exists, for you, a single real valued function for decisions (up to positive linear transformations), which we call the unconditional value $v(.)$, or your utility function, such that you will choose decision d_2 and not decision d_1 if and only if $v(d_1) \leq v(d_2)$. Thus, every decision problem can be reduced, in principle, to the decision influence diagram in Figure B.1.

The value function is unconditional in the sense that it does not depend on the state of unknown quantities. We say that the value function is conditional when it does depend on unknown quantities.

Self-consistency [H. Rubin (1987)]. Let A be the set of all possible actions that you may consider in your lifetime. Let a subset of actions $D \subset A$ correspond to a particular "decision problem" that you will face at some point in time. Let

$C(D) \subset H(D)$ be those actions that you will choose. (The operative phrase is "will choose" and the possibility that there may be more than one implies that you are indifferent between actions in $C(D)$.) $H(D)$ is the convex hull of D, i.e., the set D plus all possible randomized decisions based on D. Of course you are indifferent between choosing any particular action in $H(D)$. In practice we would simply choose an action in a randomized strategy that has the largest attached probability; i.e., we would not use randomized strategies in practice. However, it is convenient to have this added generality for the definition of self-consistency.

Call the set function $C(\bullet)$ your "choice set function" for decision problems. Self-consistency will be defined in terms of specific properties of this choice set function that you will employ for all decision problems. A person who always makes decisions by using a choice function that satisfies those properties is said to be "rational."

(1) The restriction of choices axiom. Suppose that for decision problem D you would choose the choice set $C(D)$. Now suppose for some reason that the decision problem is reduced to $D' \subset D$ but everything else is the same. If any of the old choices in $C(D)$ are in $H(D')$, then $C(D')$ would be exactly those choices from $C(D)$ that are available for decision problem D'.

(2) The weak conditionality axiom. If several decision problems are presented at random with known probabilities, the choices for each problem should be made independently of those known probabilities.

Another way of expressing this idea is through what Basu (1988) calls the "weak conditionality principle." Suppose that two experiments E_1 and E_2 are contemplated so that you may learn about some unknown quantity θ. Suppose you choose which experiment to perform by using a random mechanism; that is, with specified probability p you will perform E_1 and with probability $1 - p$ you will perform E_2. Now suppose experiment E has been determined in this way, using a random mechanism, and the results from experiment E, say, x, are available. By the weak conditionality principle, any decision about θ based on observing x should not depend on the known probability p used to determine the experiment E; i.e., knowing that it was a randomized experiment does not provide any information additional to that provided by x. What could have happened but did not should not influence your decision!

(3) The continuity axiom (Archimedean axiom). Suppose that for the decision problem $D = \{x, y\}$ you would choose x but not y; i.e., $C(\{x, y\}) = \{x\}$ and that given decision problem $D' = \{y, z\}$ you would choose y but not z, i.e., $C(\{y, z\}) = \{y\}$. By the continuity axiom we assume that there exists a probability $p(0 < p < 1)$ such that you would be indifferent between the choice

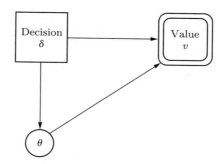

FIG. B.2.

y and a gamble, g_p, where you would choose x with probability p and z with probability $1 - p$. That is, $c(\{y, g_p\}) = \{y, g_p\}$.

In addition to the axioms above, there is an axiom stating that for every decision problem based on 1, 2, or 3 actions only, the choice set should be nonempty. This, together with the continuity axiom, is required for the construction of a utility function. In addition, H. Rubin has required two purely technical axioms to take care of the fact that we must deal with infinite sets since the use of randomization will result in conceptually infinite choice sets. He proves, using these five axioms, that if you are self-consistent in the sense of the axioms, then using your choice set function is equivalent to basing your decision on a utility function which is real valued and unique up to positive linear transformations.

Rubin's first theorem states that to base your decision on a utility function $v(d)$ means to prefer d' to d'' if and only if $v(d') > v(d'')$.

The principle of maximizing unconditional utility. To determine the solution to a decision problem in a self-consistent manner, we need only determine the unconditional utility function, a real-valued function of decisions contemplated, and take that decision (or those decisions) that maximize this function. The basic mathematical operation is that of maximizing unconditional utility with respect to possible decisions. Calculating the unconditional utility or value function is the main mathematical problem of decision analysis.

Eliminating probabilistic nodes: Calculating unconditional utility. Let θ be an unknown quantity related to a decision problem D. Let $v(d, \theta)$ be the conditional value function for our decision problem, conditional on θ. See Figure B.2.

We can eliminate node θ, obtaining the influence diagram in Figure B.1 where node v now has unconditional value

$$v(d) = \int v(d, \theta) p(\theta \mid d) d\theta$$

and where $p(\bullet \mid d)$ is the probability function attached to node θ. The math-

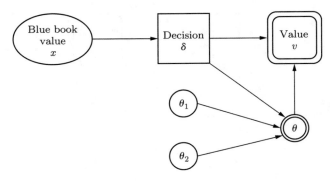

FIG. B.3.

ematical justification for this elimination operation depends on the *sure thing principle* or axiom 0 in Rubin's (1987) paper.

The sure thing principle (axiom 0).   If we prefer d_1 to d_2 given $\theta = \theta_1$ and we also prefer d_1 to d_2 given $\theta = \theta_2$, then we will prefer d_1 to d_2 when we only know that either $\theta = \theta_1$ or $\theta = \theta_2$.

As we have represented our decision problem in Figure B.2, the sure thing principle does not apply since θ depends on the decision contemplated. The same situation held in our example of selling a car in Example 10.3.3. In order that the sure thing principle apply we need to represent our decision problem by an influence diagram where decision nodes do not have adjacent successor nodes that are probabilistic. This can be done by adding a deterministic node as we have done for the selling a car example (see Figure B.3).

Example (Selling a car (*Continued*)).   As in Example 10.3.3, there are two decisions. Namely, d_1: paint your car before selling and d_2: sell without painting. In Figure B.3 we have modified Figure 10.3.2 to include a deterministic node and additional probabilistic nodes corresponding to the selling price tomorrow of cars like yours. In Figure B.3 the random node θ_1 (θ_2) is now the selling price tomorrow of cars like yours which have (have not) been painted. To apply the sure thing principle, the "state of nature" must correspond to something you can think about apart from any decisions that you are contemplating. Thus you do not think about the selling price of your car tomorrow after you make your decision but, instead, the selling price tomorrow of cars like yours corresponding to the two decisions. The point is very subtle but important.

DEFINITION (the Howard canonical form).   *A decision influence diagram is said to be in "Howard canonical form" if there are no arcs from decision nodes to probabilistic nodes. A decision influence diagram can always be put in "Howard canonical form". R. A. Howard emphasized this point.*

References

BARLOW, R. E., 1979, *Geometry of the total time on test transform*, Naval Res. Logist. Quarterly, 26, pp. 393–402.

BARLOW, R. E. and R. CAMPO, 1975, *Total time on test processes and applications to failure data analysis*, in Reliability and Fault Tree Analysis, R. E. Barlow, J. B. Fussell, and N. Singpurwalla, eds., SIAM, Philadelphia.

BARLOW, R. E. AND K. D. HEIDTMANN, 1984, *Computing k-out-of-n system reliability*, IEEE Trans. Reliability, R-33, pp. 322–323.

BARLOW, R. E. AND S. IYER, 1988, *Computational complexity of coherent systems and the reliability polynomial*, Probab. Engrg. Inform. Sci., 2, pp. 461–469.

BARLOW, R. E. AND M. B. MENDEL, 1992, *De Finetti-type representations for life distributions*, J. Amer. Statist. Assoc., 87, pp. 1116–1122.

BARLOW, R. E. AND C. A. PEREIRA, 1991, *Conditional independence and probabilistic influence diagrams*, in Topics in Statistical Dependence, IMS Lecture Notes-Monograph Series, V. 16, H. W. Block, A. R. Sampson, and T. H. Savits, eds., Institute of Mathematical Statistics, Hayward, CA, pp. 19–33.

BARLOW, R. E. AND C. A. PEREIRA, 1993, *Influence diagrams and decision modeling*, in Reliability and Decision Making, R. E. Barlow, C. Clarotti, and F. Spizzichino, eds., Chapman and Hall, New York, pp. 87–99.

BARLOW, R. E. AND F. PROSCHAN, 1973, *Availability theory for multicomponent systems*, in Multivariate Analysis III, Academic Press, Inc., New York, pp. 319–335.

BARLOW, R. E. AND F. PROSCHAN, 1975, *Importance of system components and fault tree events*, Stochastic Process. Appl., 3, pp. 153–173.

BARLOW, R. E. AND F. PROSCHAN, 1981, *Statistical Theory of Reliability and Life Testing*, To Begin With, c/o Gordon Pledger, 1142 Hornell Drive, Silver Spring, MD 20904.

BARLOW, R. E. AND F. PROSCHAN, 1985, *Inference for the exponential life distribution*, in Theory of Reliability, A. Serra and R. E. Barlow, eds., Soc. Italiana di Fisica, Bologna, Italy.

BARLOW, R. E. AND F. PROSCHAN, 1988, *Life distribution models and incomplete data*, in Handbook of Statistics, Vol. 7, P. R. Krishnaiah and C. R. Rao, eds., pp. 225–249.

BARLOW, R. E. AND F. PROSCHAN, 1996, *Mathematical Theory of Reliability*, SIAM, Philadelphia, PA.

BARLOW, R. E. AND X. ZHANG, 1986. *A critique of Deming's discussion of acceptance sampling procedures*, in Reliability and Quality Control, A. P. Basu, ed., Elsevier–North Holland, Amsterdam, pp. 9–19.

BARLOW, R. E. AND X. ZHANG, 1987, *Bayesian analysis of inspection sampling procedures discussed by Deming*, J. Statist. Plann. Inference, 16, pp. 285–296.

BASU, D., 1988, *Statistical Information and Likelihood*, Lecture Notes in Statistics 45, Springer-Verlag, New York.

BASU, D. AND C. A. PEREIRA, 1982, *On the bayesian analysis of categorical data: The problem of nonresponse*, J. Statist. Plann. Inference, 6, pp. 345–362.

BAZOVSKY, I., N. R. MACFARLANE, AND R. WUNDERMAN, 1962, *Study of Maintenance Cost Optimization and Reliability of Shipboard Machinery*, Report for ONR contract Nonr-37400, United Control, Seattle, WA.

BERGMAN, B., 1977, *Crossings in the total time on test plot*, Scand. J. Statist., 4, pp. 171–177.

BICKEL, P. J. AND D. BLACKWELL, 1967, *A note on Bayes estimates*, Ann. Math. Statist., 38, pp. 1907–1911.

BIRNBAUM, Z. W., J. D. ESARY, AND A. W. MARSHALL, 1966, *Stochastic characterization of wearout for components and systems*, Ann. Math. Statist., 37, pp. 816–825.

BIRNBAUM, Z. W., J. D. ESARY, AND S. C. SAUNDERS, 1961, *Multi-component systems and structures and their reliability*, Technometrics, 3, pp. 55–77.

CHANG, M. K. AND A. SATYANARAYANA, 1983, *Network reliability and the factoring theorem*, Networks, 13, pp. 107–120.

COLBOURN, C. J., 1987, *The Combinatorics of Network Reliability*, Oxford University Press, Oxford, England.

DAWID, A. P., 1979, *Conditional independence in statistical theory*, J. Roy. Statist. Soc. Ser. B, 41, pp. 1–31.

DE FINETTI, B., 1937, *Foresight: Its logical laws, its subjective sources*, Ann Inst. H. Poincaré, pp. 1–68; in Studies in Subjective Probability, 2nd ed., H. E. Kyburg, Jr. and H. E. Smokler, eds., Robert E. Krieger Pub. Co., Huntington, NY, 1980, pp. 53–118 (in English).

DE FINETTI, B., 1954, *Media di decisioni e media di opinioni*, Bull. Inst. Inter. Statist., 34, pp. 144–157; in Induction and Probability, a cura di Paola Monari e Daniela Cocchi, Cooperativa Libraria Universitaria Editrice Bologn, 40126 Bologna- Via Marsala 24, 1993, pp. 421–438 (in English).

DE FINETTI, B., 1970 (reprinted 1978), *Theory of Probability*, Vols. I and II, J. Wiley & Sons, New York.

DEGROOT, M. H., 1970, *Optimal Statistical Decisions*, McGraw–Hill, New York.

DEMING, W. E., 1986, *Out of the Crisis*, MIT Center for Advanced Engineering Study, Cambridge, MA.

EPSTEIN, B. AND M. SOBEL, 1953, *Life Testing*, J. Amer. Statist. Assoc., 48, pp. 486–502.

FEYNMAN, R., 1964 (sixth printing 1977), *The Feynman Lectures on Physics*, Vols. I and II, Addison–Wesley, Reading, MA.

FOX, B., 1966, *Age replacement with discounting*, Oper. Res., 14, pp. 533–537.

FUSSELL, J. B. AND W. E. VESELY, 1972, *A new methodology for obtaining cut sets for fault trees*, American Nuclear Trans., 15, pp. 262–263.

GERTSBAKH, I. B., 1989, *Statistical Reliability Theory*, Marcel Dekker, New York.

GNEDENKO, B. V., 1943, *Sur la distribution limite du terme maximum d'une série aléatorire*, Ann. of Math., 44, pp. 423–453.

HAHN, G. J. AND S. S. SHAPIRO, 1974, *Statistical Models in Engineering*, John Wiley, New York.

HESSELAGER, O., M. B. MENDEL, AND J. F. SHORTLE, 1995, *When to Use the Poisson Model (and when not to)*, manuscript.

HOFFMAN, O. AND G. SACHS, 1953, *Introduction to the Theory of Plasticity for Engineers*, McGraw-Hill, New York.

HOWARD, R. A., 1990, *From influence to relevance to knowledge*, in Belief Nets and Decision Analysis, R. M. Oliver and J. Q. Smith, eds., John Wiley, New York, pp. 3–23.

HOWARD, R. A. AND J. E. MATHESON, 1984, *Influence Diagrams*, in The Principles and Applications of Decision Analysis, Vol. II, R. A. Howard and J. E. Matheson, eds., Strategic Decisions Group, Menlo Park, CA.

HØYLAND, A. AND M. RAUSAND, 1994, *System Reliability Theory*, John Wiley, New York.

LEITCH, R. D., 1995, *Reliability Analysis for Engineers*, Oxford University Press, Oxford, England.

LINDLEY, D. V., 1985, *Making Decisions*, 2nd ed., John Wiley, New York.

LINDLEY, D. V. AND M. R. NOVICK, 1981, *The role of exchangeability in inference*, Ann. Statist., 9, pp. 45–58.

LINDQUIST, E. S., 1994, *Strength of materials and the Weibull distribution*, Probab. Engrg. Mech., 9, pp. 191–194.

MENDEL, M. B., 1989, *Development of Bayesian Parametric Theory with Applications to Control*, Ph.D. thesis, MIT, Cambridge, MA.

MOORE, E. F., AND C. E. SHANNON, 1956, *Reliable circuits using less reliable relays*, J. Franklin Inst., 262, Part I, pp. 191–208, and 262, Part II, pp. 281–297.

PARK, C. S., 1997, *Contemporary Engineering Economics*, 2nd ed., Addison–Wesley, Menlo Park, CA.

PEREIRA, C. A., 1990, *Influence diagrams and medical diagnosis*, in Influence Diagrams, Belief Nets and Decision Analysis, R. M. Oliver and J. Q. Smith, eds., John Wiley, New York, pp. 351–358.

RAI, S. AND D. P. AGRAWAL, 1990, *Distributed Computing Network Reliability*, IEEE Computer Society Press, Los Alamitos, CA.

ROSS, S. M., 1975, *On the Calculation of Asymptotic System Reliability Characteristics*, in Reliability and Fault Tree Analysis, SIAM, Philadelphia.

RUBIN, H., 1987, *A weak system of axioms for "rational" behavior and the non-separability of utility from prior*, Statist. Decisions, 5, pp. 47–58.

SHACHTER, R., 1986, *Evaluating influence diagrams*, Oper. Res., 34, pp. 871–882.

SHORTLE, J. AND M. B. MENDEL, 1994, *The Geometry of Bayesian Inference*, Bayesian Statistics 5, J. M. Bernardo, A. P. Dawid, J. Q. Smith, eds., Oxford University Press, Oxford, England.

SHORTLE, J. AND M. B. MENDEL, 1996, *Predicting dynamic imbalance in rotors*, Probab. Engrg. Mech., 11, pp. 31–35.

TSAI, P., 1994, *Probability Applications in Engineering*, Ph.D. thesis, University of California, Berkeley, CA.

VAN NOORTWIJK, J. M., 1996, *Optimal Maintenance Decisions for Hydraulic Structures under Isotropic Deterioration*, Ph.D. thesis, Delft University of Technology.

VESELY, W. E., F. F. GOLDBERG, N. H. ROBERTS, AND D. F. HAASL, 1981, *Fault Tree Handbook*, NUREG-0492, GPO Sales Program, Division of Technical Inf. and Doc. Control, U.S. Nuclear Regulatory Commission, Wash. D.C. 20555.

VINOGRADOV, O., 1991, *Introduction to Mechanical Reliability: A Designer's Approach*, Hemisphere Pub. Corp., New York.

WEIBULL, W., 1939, *A statistical theory of the strength of materials*, Ingeniorsvetenskapsakademiens Handlingar, 151, pp. 1–45.

Index

Boldface page numbers indicate location of defined term.